Optimal Mean Reversion Trading

Mathematical Analysis and Practical Applications

Modern Trends in Financial Engineering

Series Editor: Tim Leung *(Columbia University, USA)*

Published:

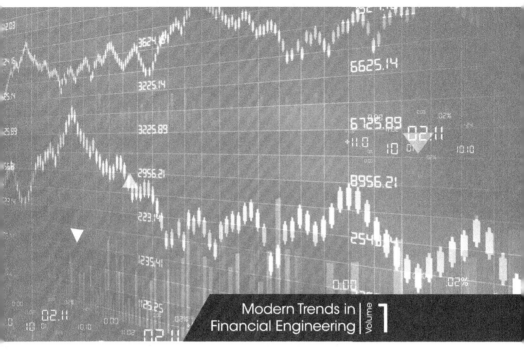

Modern Trends in Financial Engineering | Volume 1

Optimal Mean Reversion Trading

Mathematical Analysis and Practical Applications

Tim Leung
Xin Li

Columbia University, USA

 **World Scientific**

NEW JERSEY · LONDON · SINGAPORE · BEIJING · SHANGHAI · HONG KONG · TAIPEI · CHENNAI · TOKYO

Published by

World Scientific Publishing Co. Pte. Ltd.
5 Toh Tuck Link, Singapore 596224
USA office: 27 Warren Street, Suite 401-402, Hackensack, NJ 07601
UK office: 57 Shelton Street, Covent Garden, London WC2H 9HE

Library of Congress Cataloging-in-Publication Data
Names: Leung, Tim (Professor of industrial engineering)
Title: Modern trends in financial engineering : optimal mean reversion trading mathematical
    analysis and practical applications / by Tim Leung, Xin Li.
Description: New Jersey : World Scientific, 2015. | Includes bibliographical references and index.
Identifiers: LCCN 2015034540 | ISBN 9789814725910 (hardcover)
Subjects: LCSH: Financial engineering. | Stocks--Mathematical models. | Investment analysis--
    Data processing. | Exchange traded funds. | Ornstein-Uhlenbeck process.
Classification: LCC HG176.7 .L478 2015 | DDC 332.63/2042--dc23
LC record available at http://lccn.loc.gov/2015034540

British Library Cataloguing-in-Publication Data
A catalogue record for this book is available from the British Library.

In-house Editor: Li Hongyan

Typeset by Stallion Press
Email: enquiries@stallionpress.com

Printed in Singapore

Preface

This book provides a systematic study on the optimal timing of trades in markets with mean-reverting price dynamics. We present a financial engineering approach that distills the core mathematical questions from different trading problems, and also incorporates the practical aspects of trading, such as model estimation, risk premia, risk constraints, and transaction costs, into our analysis. Self-contained and organized, the book not only discusses the mathematical framework and analytical results for the financial problems, but also gives formulas and numerical tools for practical implementation. A wide array of real-world applications are discussed, such as pairs trading of exchange-traded funds, dynamic portfolio of futures on commodities or volatility indices, and liquidation of options or credit risk derivatives.

A core element of our mathematical approach is the theory of optimal stopping. For a number of the trading problems discussed herein, the optimal strategies are represented by the solutions to the corresponding optimal single/multiple stopping problems. This also leads to the analytical and numerical studies of the associated variational inequalities or free boundary problems. We provide an overview of our methodology and chapter outlines in the Introduction.

Our objective is to design the book so that it can be useful for doctoral and masters students, advanced undergraduates, and researchers in financial engineering/mathematics, especially those who specialize in algorithmic trading, or have interest in trading exchange-traded funds, commodities, volatility, and credit risk, and related derivatives. For practitioners, we provide formulas for instant strategy implementation, propose new trading strategies with mathematical justification, as well as quantitative enhancement for some existing heuristic trading strategies.

The authors would like to express their gratitude to several people who have helped make this book possible. The research conducted in collaboration with Mike Ludkovski, Peng Liu, Yoshihiro Shirai, and Zheng Wang has contributed significantly to various chapters of the book. We also appreciate the Ph.D. and master's students at Columbia University, for participating in exploratory projects related to pricing and trading under mean reversion. Our research is partially supported by NSF grant DMS-0908295. We have also benefited from the helpful remarks and suggestions by Rene Carmona, Vicky Henderson, Sebastian Jaimungal, Dimitry Kramkov, Kiseop Lee, Alexander Novikov, Mariana Olvera-Cravioto, Victor de la Pena, Scott Robertson, Neofytos Rodosthenous, Steven Shreve, Ronnie Sircar, Stathis Tompaidis, Kazutoshi Yamazaki, David Yao, Thaleia Zariphopoulou, and Hongzhong Zhang.

Last, but certainly not least, we thank Rochelle Kronzek and Max Phua of World Scientific for encouraging us to pursue this book project.

T. Leung

July 30, 2015

Contents

Chapter 1

Introduction

In the financial markets, it has been widely observed that many asset prices exhibit mean reversion, including commodities, foreign exchange rates, volatility indices, as well as US and global equities.[1] Mean-reverting processes are also used to model the dynamics of bond prices, interest rate, and default risk. In order to visualize a mean-reverting price path, we illustrate in Figure 1.1(a) the historical prices of an exchange-traded fund (ETF), the Vanguard Short-Term Bond ETF (BSV), from June 12, 2014 to June 11, 2015. This ETF is designed to track bond prices with short maturities, and is traded liquidly on NYSE and other exchanges. As another example, Figure 1.1(b) shows the time series of CBOE Volatility Index (VIX) from June 12, 2014 to June 11, 2015. Although the volatility index is not traded, investors can gain exposure to it by trading futures, options, or exchange-traded notes (ETNs) designed to track the index.[2]

In industry, hedge fund managers and investors often attempt to construct mean-reverting prices by simultaneously taking positions in two highly correlated or co-moving assets. The advent of exchange-traded funds has further facilitated this *pairs trading* approach since some ETFs are designed to track identical or similar indexes and assets. Empirical studies have found that the spreads between commodity ETFs, such as physical gold and gold equity ETFs, are mean-reverting, and such price behavior has been used for statistical arbitrage.[3]

---

[1]See Schwartz (1997) for commodities, Engel and Hamilton (1989); Anthony and Mac-Donald (1998); Larsen and Sørensen (2007) for foreign exchange rates, Metcalf and Hassett (1995); Bessembinder *et al.* (1995); Casassus and Collin-Dufresne (2005) for volatility indices, and Poterba and Summers (1988); Malliaropulos and Priestley (1999); Balvers *et al.* (2000); Gropp (2004) for US and global equities.

[2]For more details, see Section 5.6 below.

[3]See Triantafyllopoulos and Montana (2011); Dunis *et al.* (2013), and Section 2.1 below.

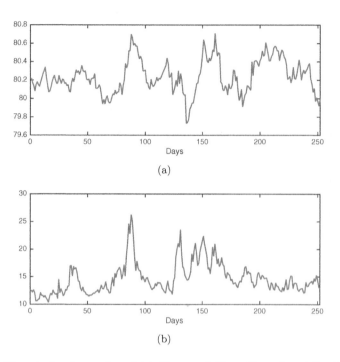

Fig. 1.1   Historical price paths of (a) Vanguard Short-Term Bond ETF (BSV) and (b) CBOE Volatility Index (VIX), respectively, from June 12, 2014 to June 11, 2015.

On the other hand, one important problem commonly faced by individual and institutional investors is to determine when to open and close a position. While observing the prevailing market prices, a speculative investor can choose to enter the market immediately or wait for a future opportunity. After completing the first trade, the investor will need to decide when is the best to close the position. This motivates the investigation of the optimal sequential timing of trades.

Naturally, the optimal sequence of trading times depend on the price dynamics of the risky asset. For instance, if the price process is a super/submartingale, that is, decreasing/increasing on average, then the investor who seeks to maximize the expected liquidation value will either sell immediately or wait forever. Such a trivial timing arises when the underlying price follows a geometric Brownian motion (see Example 2.1 below).

In this book, we study the optimal timing of trades for assets or portfolios that have mean-reverting dynamics. Specifically, we provide detailed mathematical analysis and implementation methods for various trading problems mainly under three important mean-reverting models: Ornstein-Uhlenbeck (OU), exponential Ornstein-Uhlenbeck (XOU), and Cox-Ingersoll-Ross (CIR) models. Due to their tractability and interpretability, these models are widely used in practice for describing and estimating mean reversion in asset prices. Therefore, the objective of this book is to introduce optimality criteria and discuss solution methods for an array of trading problems, and we focus on developing optimal strategies that maximize expected returns with controlled/limited risks.

1.1 Chapter Outline

In Chapter 2, we study the optimal timing of trades subject to transaction costs under the OU model. We motivate through a pairs trading example where the resulting optimized portfolio value admits an OU process. The trading strategies are implemented for the application of trading a pair of ETFs with similar underlying assets. Mathematically, our formulation leads to an *optimal double stopping* problem that gives the optimal entry and exit decision rules. We obtain analytic solutions for both the entry and exit problems. In addition, we incorporate a stop-loss constraint to our trading problem. We find that a higher stop-loss level induces the investor to voluntarily liquidate earlier at a lower take-profit level. Moreover, the entry region is characterized by a bounded price interval that lies strictly above stop-loss level. In other words, it is optimal to wait if the current price is too high or too close to the lower stop-loss level. This is intuitive since entering the market close to stop-loss implies a high chance of exiting at a loss afterwards. As a result, the delay region (complement of the entry region) is *disconnected*. Furthermore, we show that optimal liquidation level decreases with the stop-loss level until they coincide, in which case immediate liquidation is optimal at all price levels.

To incorporate mean-reversion for positive price processes, one popular choice for pricing and investment applications is the exponential OU model, as proposed by Schwartz (1997) for commodity prices, due to its analytical tractability. It also serves as the building block of more sophisticated mean-reverting models. In Chapter 3, we study the optimal timing of trades under the XOU model. We consider the *optimal double stopping* problem, as well as a different but related formulation. In the second formulation, the investor is assumed to enter and exit the market infinitely many times with transaction costs. This gives rise to an *optimal switching* problem. We analytically derive the non-trivial entry and exit timing strategies. Under both approaches, it is optimal to sell when the asset price is sufficiently high, though at different levels. As for entry timing, we find that, under some conditions, it is optimal for the investor not to enter the market at all when facing the optimal switching problem. In this case for the investor who has a long position, the optimal switching problem reduces into an optimal stopping problem, where the optimal liquidation level is identical to that of the optimal double stopping problem. Otherwise, the optimal entry timing strategies for the double stopping and switching problem are described by the underlying's first passage time to an interval that lies above level zero. In other words, the continuation region for entry is *disconnected* of the form $(0, A) \cup (B, +\infty)$, with critical price levels A and B (see Theorems 3.4 and 3.7 below). This means that the investor generally enters when the price is low, but may find it optimal to wait if the current price is too close to zero. We find that this phenomenon is a distinct consequence due to fixed transaction costs under the XOU model.

In Chapter 4, we turn to the trading problems when the asset follows the CIR process. The CIR process has been widely used as the model for interest rate, volatility, commodity, and energy prices.[4] The main focus of the chapter is the analytical derivation of the non-trivial optimal entry and exit timing strategies and the associated value functions. Under both double stopping and switching approaches, it is optimal to exit when the process value is sufficiently high, though at different levels. As for entry timing, we find the necessary and sufficient conditions whereby it is optimal not to enter at all when facing the optimal switching problem. In this case, the optimal switching problem in fact reduces to an optimal single stopping problem, where the optimal stopping level is identical to that of the optimal double stopping problem.

A typical solution approach for optimal stopping problems driven by diffusion involves the analytical and numerical studies of the associated free boundary problems or variational inequalities (VIs); see, for example, Bensoussan and Lions (1982), Øksendal (2003), and Sun (1992). For our double optimal stopping problem, this method would determine the value functions from a pair of VIs and require regularity conditions to guarantee that the solutions to the VIs indeed correspond to the optimal stopping problems. As noted by Dayanik (2008), "the variational methods become

---

[4]See, for example, Cox *et al.* (1985); Ewald and Wang (2010); Heston (1993); Ribeiro and Hodges (2004), among others.

challenging when the form of the reward function and/or the dynamics of the diffusion obscure the shape of the optimal continuation region." In our optimal entry timing problem, the reward function involves the value function from the exit timing problem, which is not monotone and can be positive and negative.

In contrast to the variational inequality approach, our proposed methodology for Chapters 2 to 4 applies probabilistic arguments to analytically characterize the optimal stopping value functions as the smallest concave majorant of the corresponding reward function. A key feature of this approach is that it allows us to directly construct the value function, without *a priori* finding a candidate value function or imposing conditions on the stopping and delay (continuation) regions, such as whether they are connected or not. In other words, our method will derive the structure of the stopping and delay regions as an output. Having solved the optimal double stopping problem, we determine the optimal structures of the buy/sell/wait regions. We then apply this to infer a similar solution structure for the optimal switching problem and verify using the variational inequalities.

Chapters 5 to 7 are dedicated to trading of financial derivatives, namely, futures, options, and credit derivatives, respectively. Started as contracts for the delivery of agricultural products decades ago, futures are now one of the most common form of financial derivatives, and there are a high number of tradable commodities, including agricultural products, livestocks, precious metals, oil and gas, as well as other underlyings such as interest rates, currency, equity and volatility indices. Each futures contract stipulates the buyer to purchase (seller to sell) a fixed quantity of a commodity at a fixed price to be paid for on a pre-specified future date. Many futures require physical delivery of the commodity, but some, like the VIX futures, are settled in cash. In Chapter 5, we discuss the pricing of futures, explore the timing options embedded in futures trading, and develop optimal dynamic speculative strategies for market entry and exit. Focusing on the applications to commodity and volatility futures, we analyze these problems under mean-reverting spot price dynamics.

For decades, options have been widely used as a tool for investment and risk management. As of 2012, the daily market notional for S&P 500 options is more than US$90 billion and the average daily volume has grown rapidly from 119,808 in 2002 to 839,108 as of Jan 2013.[5] Empirical studies on options returns often assume that the options are held to maturity (see

---

Broadie *et al.* (2009) and references therein). In a liquid options market, such as the S&P 500 index, VIX, or gold options markets, there is an intrinsic timing flexibility to liquidate the position through the market prior to expiry. This leads us to investigate the optimal time to liquidate an option position. In Chapter 6, we propose a risk-adjusted optimal stopping framework to address this problem for a variety of options under different underlying price dynamics.

In addition to maximizing the expected discounted market value to be received from option sale, we incorporate a risk penalty that accounts for adverse price movements till the liquidation time. Specifically, we measure the associated risk in terms of the realized shortfall of the option position, and thus introduce the trade-off between risk and return for every liquidation timing strategy. Under a general diffusion model for the underlying stock price, we formulate an optimal stopping problem that includes an integral penalization term. To this end, we define and apply the concept of *optimal liquidation premium* which represents the additional value from optimally waiting to sell, as opposed to immediate liquidation. As it turns out, it is optimal for the option holder to sell as soon as this premium vanishes. This observation leads to a number of useful mathematical characterizations and financial interpretations of the optimal liquidation strategies for various positions.

Lastly, in Chapter 7 we propose a new approach to tackle the optimal liquidation problem for credit derivatives. The first step is to understand how the market compensates investors for bearing credit risk. We examine analytically the structure of default risk premia inferred from the market prices of corporate bonds, credit default swaps, and multi-name credit derivatives. We identify the risk premium components, namely, the *mark-to-market risk premium* that accounts for the fluctuations in default risk, as well as the *event risk premium* (or jump-to-default risk premium) that compensates for the uncertain timing of the default event. Our approach is to first provide a general mathematical framework for price discrepancy between the market and investors under an intensity-based credit risk model. Then, we derive and analyze the optimal stopping problem corresponding to the liquidation of credit derivatives under price discrepancy.

In order to measure the benefit of optimally timing to sell as opposed to immediate liquidation, we define and quantify the so-called *delayed liquidation premium*. We analyze the scenarios where immediate or delayed liquidation is optimal. Moreover, through its probabilistic representation, the delayed liquidation premium reveals the roles of risk premia in the liquidation timing. We investigate the investor's liquidation timing for various credit derivatives, including defaultable bonds, CDSs, as well as, multi-name credit derivatives, in markets where the default intensity and interest rate processes are mean-reverting. The impact of price discrepancy is revealed through a series of numerical examples illustrating the trader's optimal liquidation strategies.

1.2 Related Studies

In the context of pairs trading, a number of studies have also considered market timing strategy with two price levels. For example, Gatev *et al.* (2006) study the historical returns from the buy-low-sell-high strategy where the entry/exit levels are set as ±1 standard deviation from the long-run mean. Similarly, Avellaneda and Lee (2010) consider starting and ending a pairs trade based on the spread's distance from its mean. In Elliott *et al.* (2005), the market entry timing is modeled by the first passage time of an OU process, followed by an exit at a fixed finite horizon. In comparison, rather than assigning *ad hoc* price levels or fixed trading times, our approach in Chapter 2, adapted from Leung and Li (2015), generates the entry and exit thresholds as solutions of an optimal double stopping problem. Considering an exponential OU asset price with zero log mean, Bertram (2010) numerically computes the optimal enter and exit levels that maximize the expected return per unit time. Other timing strategies adopted by practitioners have been discussed in Vidyamurthy (2004). Song *et al.* (2009) and Song and Zhang (2013) study the optimal switching problem with stop-loss under the OU price dynamics. In their recent book, Cartea *et al.* (2015) also study the pairs trading problem that allows the investor to enter the market by longing one asset and shorting the other, *or* taking the opposite position, thus resulting in a two-sided market entry strategy.

In Chapters 3 and 4, we consider optimal double stopping and optimal switching problems under exponential OU and CIR models with fixed transaction costs. These two chapters are based on Leung *et al.* (2015) and Leung *et al.* (2014). In particular, the optimal entry timing with fixed transaction costs is characteristically different from that with slippage (see Czichowsky *et al.* (2015); Kong and Zhang (2010); Zhang and Zhang (2008)). Zervos *et al.* (2013) consider an optimal switching problem with fixed transaction costs under some time-homogeneous diffusions, including the GBM, mean-reverting CEV underlying, but their results are not applicable to the exponential OU model.

On the other hand, the related problem of constructing portfolios and hedging with mean-reverting asset prices has been studied. For example, Benth and Karlsen (2005) study the utility maximization problem that involves dynamically trading an exponential OU underlying asset. Jurek and Yang (2007) analyze a finite-horizon portfolio optimization problem with an OU asset subject to the power utility and Epstein-Zin recursive utility. Chiu and Wong (2012) consider the dynamic trading of co-integrated assets with a mean-variance criterion. Tourin and Yan (2013) derive the dynamic trading strategy for two co-integrated stocks in order to maximize the expected terminal utility of wealth over a fixed horizon. They simplify the associated Hamilton-Jacobi-Bellman equation and obtain a closed-form solution. In the stochastic control approach, incorporating transaction costs and stop-loss exit can potentially limit model tractability and is not implemented in these studies.

In terms of methodology for Chapters 2–4, Dynkin and Yushkevich (1969) analyze the concave characterization of excessive functions for a standard Brownian motion. Dayanik (2008) and Dayanik and Karatzas (2003) apply this idea to study the optimal single stopping of a one-dimensional diffusion. Alvarez (2003) discusses the conditions for the convexity of an r-excessive mapping under a linear, time-homogeneous and regular diffusion process. Menaldi *et al.* (1996) study an optimal starting-stopping problem for general Markov processes, and provide the mathematical characterization of the value functions. In this regard, we contribute to this line of work by solving a number of optimal double stopping problems under the OU, XOU, and CIR models, and incorporating a stop-loss exit under the OU model.

In Chapter 5, we study the pricing and trading of futures under a mean-reverting spot price model. This is most relevant to the futures markets on volatility indices and commodities. For prior work on the valuation of volatility futures and options, we refer to Grübichler and Longstaff (1996); Lin and Chang (2009); Mencía and Sentana (2013); Sircar and Papanicolaou (2014); Zhang and Zhu (2006). A number of mean-reverting spot price models have been proposed for the pricing of commodity futures; see Schwartz (1997) and references therein. On the trading of futures, Brennan and Schwartz (1990) and Dai *et al.* (2011) investigate optimal timing to capture the arbitrage opportunity embedded in the spread between the futures and the spot prices, and model the stochastic spread by a Brownian bridge that is pinned to level zero at maturity. Similarly, Kanamura *et al.* (2010) model the spread between energy futures prices as an OU process.

These two models can be considered as a reduced form approach whereby the spread is modeled directly without regard to the spot price dynamics or term structure of the futures. In contrast, we model for our trading problems the fundamental source of randomness, that is, the spot price, that drives all futures prices.

Our path-dependent risk penalization model in Chapter 6 can also be viewed as an alternative way to incorporate the investor's risk sensitivity in option liquidation/exercise timing problems, as compared to the utility maximization/indifference pricing approach (see Henderson and Hobson (2011); Leung and Ludkovski (2012); Leung *et al.* (2012)). As is well known, the concept of risk measures based on shortfall risk has been applied to many portfolio optimization problems; see Artzner *et al.* (1999); Föllmer and Schied (2002); Föllmer and Schied (2004); İlhan *et al.* (2005); Rockafellar and Uryasev (2000), and references therein. Our model applies this idea to options trading as a path-penalty associated with each liquidation strategy. As a variation of the shortfall we also introduce a risk penalty based on the quadratic variation of option price process. Through examining the optimal liquidation premium, we also compare the liquidation strategies for calls and puts under the shortfall-based and quadratic risk penalties. Forsyth *et al.* (2012) and Frei and Westray (2013) also adopt the mean-quadratic-variation as a criterion for determining the optimal stock trading strategy in the presence of price impact. The related work by Leung and Liu (2013) that discusses the timing to sell an option under the GBM model without any risk penalty is a special example of our model.

Our study on credit derivative liquidation timing in Chapter 7 is closest to Leung and Ludkovski (2011), where the concept of *delayed purchase premium* was used to analyze the optimal timing to purchase equity European and American options. In contrast, we adopt a multi-factor intensity-based default risk model for single-name credit derivatives, and a self-exciting top-down model for a credit default index swap. As a natural extension, we also investigate the optimal timing to buy and sell a credit derivative, with or without short-sale constraint, and provide numerical illustration of the the optimal buy-and-sell strategy. In the related papers, Egami *et al.* (2013) and Leung and Yamazaki (2013) incorporate the timing option to terminate a credit default swap under a structural default model where the underlying asset is driven by a Lévy process.

In Chapter 7, we also consider the connection between different risk-neutral pricing measures (or equivalent martingale measures) in incomplete markets. Well-known examples of candidate pricing measures that are consistent with the no-arbitrage principle include the minimal martingale measure (Föllmer and Schweizer (1990)), the minimal entropy martingale measure (Fritelli (2000); Fujiwara and Miyahara (2003)), and the q-optimal martingale measure (Henderson *et al.* (2005); Hobson (2004)). The investor's selection of various pricing measures may also be interpreted via marginal utility indifference valuation; see Davis (1997); Leung and Ludkovski (2012); Leung *et al.* (2012), and references therein. For many parametric credit risk models, the market pricing measures and risk premia can be calibrated given sufficient market data of credit derivatives.

In recent literature, a number of models have been proposed to incorporate the idea of mispricing into optimal investment. Cornell *et al.* (2007) study portfolio optimization based on perceived mispricing from the investor's strong belief in the stock price distribution. Ekström *et al.* (2010) investigate the optimal liquidation of a call spread when the investor's belief on the volatility differs from the implied volatility. The problem of timing to buy/sell European and American options has also been studied in Leung and Ludkovski (2011, 2012). On the other hand, the problem of optimal stock liquidation involving price impacts has been studied in Almgren (2003); Rogers and Singh (2010); Schied and Schöneborn (2009), among others.

Chapter 2

Trading Under the Ornstein-Uhlenbeck Model

Motivated by the industry practice of pairs trading, we study the optimal timing strategies for trading a mean-reverting price spread. An optimal double stopping problem is formulated to analyze the timing to start and subsequently liquidate the position subject to transaction costs. Modeling the price spread by an Ornstein-Uhlenbeck process, we apply a probabilistic methodology and rigorously derive the optimal price intervals for market entry and exit. A number of extensions are also considered, such as incorporating a stop-loss constraint, or a minimum holding period. We show that the entry region is characterized by a bounded price interval that lies strictly above the stop-loss level. As for the exit timing, a higher stop-loss level always implies a lower optimal take-profit level. Both analytical and numerical results are provided to illustrate the dependence of timing strategies on model parameters such as transaction costs and stop-loss level.

In Section 2.1, we discuss a pairs trading example with OU price spreads, and formulate the optimal trading problem. Our method of solution is presented in Section 2.3. In Section 2.4, we analytically solve the optimal double stopping problem and examine the optimal entry and exit strategies. In Section 2.5, we study the trading problem with a stop-loss constraint. In Section 2.6, we present a number of extensions. The proofs of all lemmas are provided in Section 2.7.

2.1 A Pairs Trading Example

Let us discuss a pairs trading example where we model the value of the resulting position by an OU process. The primary objective is to motivate our trading problem, rather than proposing new estimation methodologies or empirical studies on pairs trading. For related studies and more details, we refer to the seminal paper by Engle and Granger (1987), the books Hamilton (1994), Tsay (2005), and references therein.

We construct a portfolio by holding α shares of a risky asset $S^{(1)}$ and shorting β shares of another risky asset $S^{(2)}$, yielding a portfolio value $X_t^{\alpha,\beta} = \alpha S_t^{(1)} - \beta S_t^{(2)}$ at time $t \geq 0$. The pair of assets are selected to form a mean-reverting portfolio value. In addition, one can adjust the strategy (α, β) to enhance the level of mean reversion. For the purpose of testing mean reversion, only the ratio between α and β matters, so we can keep α constant while varying β without loss of generality. For every strategy (α, β), we observe the resulting portfolio values $(x_i^{\alpha,\beta})_{i=0,1,\ldots,n}$ realized over an n-day period. We then apply the method of maximum likelihood estimation (MLE) to fit the observed portfolio values to an OU process and determine the model parameters.

We fix the probability space $(\Omega, \mathcal{F}, \mathbb{P})$ with the historical probability measure $\mathbb{P}$. We consider an Ornstein-Uhlenbeck (OU) process driven by the SDE:

$$dX_t = \mu(\theta - X_t)\,dt + \sigma\,dB_t, \tag{2.1}$$

with constants $\mu, \sigma > 0$, $\theta \in \mathbb{R}$, and state space $\mathbb{R}$. Here, B is a standard Brownian motion under $\mathbb{P}$. Denote by $\mathbb{F} \equiv (\mathcal{F}_t)_{t\geq 0}$ the filtration generated by X.

Under the OU model, the conditional probability density of X_{t_i} at time t_i given $X_{t_{i-1}} = x_{i-1}$ with time increment $\Delta t = t_i - t_{i-1}$ is given by

$$f^{OU}(x_i|x_{i-1}; \theta, \mu, \sigma) = \frac{1}{\sqrt{2\pi\tilde{\sigma}^2}}\exp\left(-\frac{(x_i - x_{i-1}e^{-\mu\Delta t} - \theta(1 - e^{-\mu\Delta t}))^2}{2\tilde{\sigma}^2}\right),$$

with the constant

$$\tilde{\sigma}^2 = \sigma^2\frac{1 - e^{-2\mu\Delta t}}{2\mu}.$$

Using the observed values $(x_i^{\alpha,\beta})_{i=0,1,\ldots,n}$, we maximize the average log-likelihood defined by

$$\ell(\theta, \mu, \sigma|x_0^{\alpha,\beta}, x_1^{\alpha,\beta}, \ldots, x_n^{\alpha,\beta})$$

$$:= \frac{1}{n}\sum_{i=1}^{n}\ln f^{OU}\left(x_i^{\alpha,\beta}|x_{i-1}^{\alpha,\beta}; \theta, \mu, \sigma\right)$$

$$= -\frac{1}{2}\ln(2\pi) - \ln(\tilde{\sigma}) - \frac{1}{2n\tilde{\sigma}^2}\sum_{i=1}^{n}[x_i^{\alpha,\beta} - x_{i-1}^{\alpha,\beta}e^{-\mu\Delta t} - \theta(1 - e^{-\mu\Delta t})]^2.$$

$$\tag{2.2}$$

To express the parameter values that maximize the average log-likelihood in (2.2) , we define the followings:

$$X_x = \sum_{i=1}^{n} x_{i-1}^{\alpha,\beta},$$

$$X_y = \sum_{i=1}^{n} x_{i}^{\alpha,\beta},$$

$$X_{xx} = \sum_{i=1}^{n} (x_{i-1}^{\alpha,\beta})^2,$$

$$X_{xy} = \sum_{i=1}^{n} x_{i-1}^{\alpha,\beta} x_{i}^{\alpha,\beta},$$

$$X_{yy} = \sum_{i=1}^{n} (x_{i}^{\alpha,\beta})^2.$$

In turn, the optimal parameter estimates under the OU model are given explicitly by

$$\theta^* = \frac{X_y X_{xx} - X_x X_{xy}}{n(X_{xx} - X_{xy}) - (X_x^2 - X_x X_y)},$$

$$\mu^* = -\frac{1}{\Delta t} \ln \frac{X_{xy} - \theta^* X_x - \theta^* X_y + n(\theta^*)^2}{X_{xx} - 2\theta^* X_x + n(\theta^*)^2},$$

$$(\sigma^*)^2 = \frac{2\mu^*}{n(1 - e^{-2\mu^* \Delta t})} (X_{yy} - 2e^{-\mu^* \Delta t} X_{xy} + e^{-2\mu^* \Delta t} X_{xx}$$
$$- 2\theta^* (1 - e^{-\mu^* \Delta t})(X_y - e^{-\mu^* \Delta t} X_x) + n(\theta^*)^2(1 - e^{-\mu^* \Delta t})^2).$$

In turn, we denote by $\hat{\ell}(\theta^*, \mu^*, \sigma^*)$ the maximized average log-likelihood.

For any α, we choose the strategy (α, β^*), where

$$\beta^* = \arg\max_{\beta} \hat{\ell}(\theta^*, \mu^*, \sigma^* | x_0^{\alpha,\beta}, x_1^{\alpha,\beta}, \ldots, x_n^{\alpha,\beta}).$$

For example, suppose we invest A dollar(s) in asset $S^{(1)}$, so $\alpha = A/S_0^{(1)}$ shares is held. At the same time, we short $\beta = B/S_0^{(2)}$ shares in $S^{(2)}$, for $B/A = 0.001, 0.002, \ldots, 1$. This way, the sign of the initial portfolio value depends on the sign of the difference $A - B$, which is non-negative. Without loss of generality, we set $A = 1$.

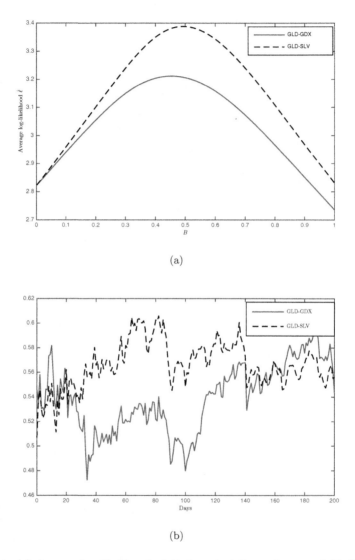

(a)

(b)

Fig. 2.1 (a) Average log-likelihood plotted against the cash amount B. (b) Historical price paths with maximum average log-likelihood. The solid line plots the portfolio price with longing $1 GLD and shorting $0.454 GDX, and the dashed line plots the portfolio price with longing $1 GLD and shorting $0.493 SLV.

In Figure 2.1, we illustrate an example based on two pairs of exchange-traded funds (ETFs), namely, the Market Vectors Gold Miners (GDX) and iShares Silver Trust (SLV) against the SPDR Gold Trust (GLD) respectively. These three liquidly traded funds aim to track the price movements of the NYSE Arca Gold Miners Index, silver, and gold bullion respectively. These ETF pairs are also used in Dunis *et al.* (2013) and Triantafyllopoulos and Montana (2011) for their statistical and empirical studies on ETF pairs trading.

Using price data from August 2011 to May 2012 ($n = 200$, $\Delta t = 1/252$), we compute and plot in Figure 2.1(a) the average log-likelihood against the cash amount B, and find that $\hat{\ell}$ is maximized at $B^* = 0.454$ (resp. 0.493) for the GLD-GDX pair (resp. GLD-SLV pair). From this MLE-optimal B^*, we obtain the strategy (α, β^*), where $\alpha = 1/S_0^{(1)}$ and $\beta^* = B^*/S_0^{(2)}$. In this example, the average log-likelihood for the GLD-SLV pair happens to dominate that for GLD-GDX, suggesting a higher degree of fit to the OU model. Figure 2.1(b) depicts the historical price paths with the strategy (α, β^*).

We summarize the estimation results in Table 2.1. For each pair, we first estimate the parameters for the OU model from empirical price data. Then, we use the estimated parameters to simulate price paths according the corresponding OU process. Based on these simulated OU paths, we perform another MLE and obtain another set of OU parameters as well as the maximum average log-likelihood $\hat{\ell}$. For the two examples, the portfolio consists of \$1 in GLD, along with either $-\$0.454$ in GDX, or $-\$0.493$ in SLV. For each pair, the second row (simulated) shows the MLE parameter estimates based on a simulated price path corresponding to the estimated parameters from the first row (empirical). As we can see, the two sets of estimation outputs (the rows named "empirical" and "simulated") are very close, suggesting the empirical price process fits well to the OU model.

Table 2.1 MLE estimates of OU model parameters using historical values of GLD-GDX and GLD-SLV portfolios from August 2011 to May 2012.

	Price	$\hat{\theta}$	$\hat{\mu}$	$\hat{\sigma}$	$\hat{\ell}$
GLD-GDX	empirical	0.5388	16.6677	0.1599	3.2117
	simulated	0.5425	14.3893	0.1727	3.1304
GLD-SLV	empirical	0.5680	33.4593	0.1384	3.3882
	simulated	0.5629	28.8548	0.1370	3.3898

2.2 Optimal Timing of Trades

When considering to trade in a market, every investor possesses a timing option to trade. The investor can choose to enter the market immediately or wait for a potentially better opportunity. This leads us to study the optimal timing to open and subsequently close the position subject to transaction costs, given that the asset price or portfolio value evolves according to an OU process.

First, suppose that the investor already has an existing position whose value process $(X_t)_{t\geq 0}$ follows (2.1). If the position is closed at some time τ, then the investor will receive the value X_τ and pay a constant transaction cost $c_s \in \mathbb{R}$. To maximize the expected discounted value, the investor solves the optimal stopping problem

$$V(x) = \sup_{\tau \in \mathcal{T}} \mathbb{E}_x\big\{e^{-r\tau}(X_\tau - c_s)\big\}, \tag{2.3}$$

where $\mathcal{T}$ denotes the set of all $\mathbb{F}$-stopping times, and $r > 0$ is the investor's subjective constant discount rate. We have also used the shorthand notation: $\mathbb{E}_x\{\cdot\} \equiv \mathbb{E}\{\cdot|X_0 = x\}$.

From the investor's viewpoint, $V(x)$ represents the expected liquidation value associated with X. On the other hand, the current price plus the transaction cost constitute the total cost to enter the trade. The investor can always choose the optimal timing to start the trade, or not to enter at all. This leads us to analyze the entry timing inherent in the trading problem. Precisely, we solve

$$J(x) = \sup_{\nu \in \mathcal{T}} \mathbb{E}_x\big\{e^{-\hat{r}\nu}(V(X_\nu) - X_\nu - c_b)\big\}, \tag{2.4}$$

with $\hat{r} > 0$, $c_b \in \mathbb{R}$. In other words, the investor seeks to maximize the expected difference between the value function $V(X_\nu)$ and the current X_ν, minus transaction cost c_b. The value function $J(x)$ represents the maximum expected value of the investment opportunity in the price process X, with transaction costs c_b and c_s incurred, respectively, at entry and exit. Mathematically, embedded in the value functions is an optimal double stopping problem.

For our analysis, the pre-entry and post-entry discount rates, $\hat{r}$ and r, can be different, as long as $0 < \hat{r} \leq r$. Moreover, the transaction costs c_b and c_s can also differ, as long as $c_s + c_b > 0$. Furthermore, since $\tau = +\infty$ and $\nu = +\infty$ are candidate stopping times for (2.3) and (2.4) respectively, the two value functions $V(x)$ and $J(x)$ are non-negative.

As extension, we can incorporate a stop-loss level of the pairs trade, that caps the maximum loss. In practice, the stop-loss level may be exogenously imposed by the manager of a trading desk. In effect, if the price X ever reaches level L prior to the investor's voluntary liquidation time, then the position will be closed immediately. The stop-loss signal is given by the first passage time

$$\tau_L := \inf\{t \geq 0 \,:\, X_t \leq L\}.$$

Therefore, we determine the entry and liquidation timing from the constrained optimal stopping problem:

$$J_L(x) = \sup_{\nu \in \mathcal{T}} \mathbb{E}_x \left\{ e^{-\hat{r}\nu} (V_L(X_\nu) - X_\nu - c_b) \right\}, \qquad (2.5)$$

$$V_L(x) = \sup_{\tau \in \mathcal{T}} \mathbb{E}_x \left\{ e^{-r(\tau \wedge \tau_L)} (X_{\tau \wedge \tau_L} - c_s) \right\}. \qquad (2.6)$$

Due to the additional timing constraint, the investor may be forced to exit early at the stop-loss level for any given liquidation level. Hence, the stop-loss constraint reduces the value functions, and precisely we deduce that $x - c_s \leq V_L(x) \leq V(x)$ and $0 \leq J_L(x) \leq J(x)$. As we will show in Sections 2.4 and 2.5, the optimal timing strategies with and without stop-loss are quite different.

Example 2.1. One well-known model for asset price is the geometric Brownian motion model:

$$dX_t = \mu X_t \, dt + \sigma X_t \, dB_t. \qquad (2.7)$$

However, the optimal timing strategies for both V and J (see (2.3) and (2.4)) with X given in (2.7) are trivial. Indeed, if $\mu > r$, then considering constant exercise times we have the inequality

$$V(x) \geq \sup_{t \geq 0} \left(\mathbb{E}_x\{e^{-rt}X_t\} - e^{-rt}c \right) \geq \sup_{t \geq 0} x e^{(\mu-r)t} - c = +\infty. \qquad (2.8)$$

Therefore, it is optimal to take $\tau = +\infty$ and the value function is infinite.

If $\mu = r$, then the value function is given by

$$V(x) = \sup_{t \geq 0} \sup_{\tau \in \mathcal{T}} \mathbb{E}_x\{e^{-r(\tau \wedge t)}(X_{\tau \wedge t} - c)\} = x - c \inf_{t \geq 0} \inf_{\tau \in \mathcal{T}} \mathbb{E}_x\{e^{-r(\tau \wedge t)}\} = x,$$

where the second equality follows from the optional sampling theorem with the fact that $(e^{-rt}X_t)_{t \geq 0}$ is a martingale. Again, the optimal value is achieved by choosing $\tau = +\infty$, but, in contrast to (2.8), $V(x)$ is finite in (2.1).

On the other hand, if $\mu < r$, then we have

$$V(x) = \begin{cases} \left(\frac{c}{\eta-1}\right)^{1-\eta} \left(\frac{x}{\eta}\right)^{\eta} & \text{if } x < b^*, \\ x - c & \text{if } x \geq b^*, \end{cases}$$

where

$$\eta = \frac{\sqrt{2r\sigma^2 + (\mu - \frac{1}{2}\sigma^2)^2} - (\mu - \frac{1}{2}\sigma^2)}{\sigma^2} \quad \text{and} \quad b^* = \frac{c\eta}{\eta-1} > c.$$

Therefore, it is optimal to liquidate as soon as X reaches level b^* from below. However, it is optimal not to enter, since $\sup_{x \in \mathbb{R}_+}(V(x) - x - c_b) \leq 0$, and thus, $J(x) = 0$. Henceforth, we study the optimal stopping problems under mean reverting dynamics.

2.3   Methodology

In this section, we disucss our method of solution. First, we denote the infinitesimal generator of the OU process X by

$$\mathcal{L} = \frac{\sigma^2}{2}\frac{d^2}{dx^2} + \mu(\theta - x)\frac{d}{dx}, \tag{2.9}$$

and recall the classical solutions of the differential equation

$$\mathcal{L}u(x) = ru(x), \tag{2.10}$$

for $x \in \mathbb{R}$, are (see e.g. p.542 of Borodin and Salminen (2002) and Prop. 2.1 of Alili *et al.* (2005)):

$$F(x) \equiv F(x;r) := \int_0^\infty u^{\frac{r}{\mu}-1} e^{\sqrt{\frac{2\mu}{\sigma^2}}(x-\theta)u - \frac{u^2}{2}} \, du, \tag{2.11}$$

$$G(x) \equiv G(x;r) := \int_0^\infty u^{\frac{r}{\mu}-1} e^{\sqrt{\frac{2\mu}{\sigma^2}}(\theta-x)u - \frac{u^2}{2}} \, du. \tag{2.12}$$

Direct differentiation yields that $F'(x) > 0$, $F''(x) > 0$, $G'(x) < 0$ and $G''(x) > 0$. Hence, we observe that both $F(x)$ and $G(x)$ are strictly positive and convex, and they are, respectively, strictly increasing and decreasing.

Define the first passage time of X to some level κ by $\tau_\kappa = \inf\{t \geq 0 : X_t = \kappa\}$. As is well known, F and G admit the probabilistic expressions (see Itō and McKean (1965) and Rogers and Williams (2000)):

$$\mathbb{E}_x\{e^{-r\tau_\kappa}\} = \begin{cases} \frac{F(x)}{F(\kappa)} & \text{if } x \leq \kappa, \\ \frac{G(x)}{G(\kappa)} & \text{if } x \geq \kappa. \end{cases}$$

A key step of our solution method involves the transformation

$$\psi(x) := \frac{F}{G}(x). \qquad (2.13)$$

Starting at any $x \in \mathbb{R}$, we denote by $\tau_a \wedge \tau_b$ the exit time from an interval $[a, b]$ with $-\infty \leq a \leq x \leq b \leq +\infty$. With the reward function $h(x) = x - c_s$, we compute the corresponding expected discounted reward:

$$\mathbb{E}_x\{e^{-r(\tau_a \wedge \tau_b)} h(X_{\tau_a \wedge \tau_b})\}$$
$$= h(a)\mathbb{E}_x\{e^{-r\tau_a}\mathbf{1}_{\{\tau_a < \tau_b\}}\} + h(b)\mathbb{E}_x\{e^{-r\tau_b}\mathbf{1}_{\{\tau_a > \tau_b\}}\} \qquad (2.14)$$
$$= h(a)\frac{F(x)G(b) - F(b)G(x)}{F(a)G(b) - F(b)G(a)} + h(b)\frac{F(a)G(x) - F(x)G(a)}{F(a)G(b) - F(b)G(a)} \qquad (2.15)$$
$$= G(x)\left[\frac{h(a)}{G(a)}\frac{\psi(b) - \psi(x)}{\psi(b) - \psi(a)} + \frac{h(b)}{G(b)}\frac{\psi(x) - \psi(a)}{\psi(b) - \psi(a)}\right]$$
$$= G(\psi^{-1}(z))\left[H(z_a)\frac{z_b - z}{z_b - z_a} + H(z_b)\frac{z - z_a}{z_b - z_a}\right], \qquad (2.16)$$

where $z_a = \psi(a)$, $z_b = \psi(b)$, and

$$H(z) := \begin{cases} \frac{h}{G} \circ \psi^{-1}(z) & \text{if } z > 0, \\ \lim\limits_{x \to -\infty} \frac{(h(x))^+}{G(x)} & \text{if } z = 0. \end{cases} \qquad (2.17)$$

The second equality (2.15) follows from the fact that $f(x) := \mathbb{E}_x\{e^{-r(\tau_a \wedge \tau_b)}\mathbf{1}_{\{\tau_a < \tau_b\}}\}$ is the unique solution to (2.10) with boundary conditions $f(a) = 1$ and $f(b) = 0$. Similar reasoning applies to the function $g(x) := \mathbb{E}_x\{e^{-r(\tau_a \wedge \tau_b)}\mathbf{1}_{\{\tau_a > \tau_b\}}\}$ with $g(a) = 0$ and $g(b) = 1$. The last equality (2.16) transforms the problem from x coordinate to $z = \psi(x)$ coordinate (see (2.13)).

The candidate optimal exit interval $[a^*, b^*]$ is determined by maximizing the expectation in (2.14). This is equivalent to maximizing (2.16) over z_a and z_b in the transformed problem. This leads to

$$W(z) := \sup_{\{z_a, z_b : z_a \leq z \leq z_b\}}\left[H(z_a)\frac{z_b - z}{z_b - z_a} + H(z_b)\frac{z - z_a}{z_b - z_a}\right]. \qquad (2.18)$$

This is the smallest concave majorant of H. Applying the definition of W to (2.16), we can express the maximal expected discounted reward as

$$G(x)W(\psi(x)) = \sup_{\{a, b : a \leq x \leq b\}} \mathbb{E}_x\{e^{-r(\tau_a \wedge \tau_b)} h(X_{\tau_a \wedge \tau_b})\}.$$

Remark 2.2. If $a = -\infty$, then we have $\tau_a = +\infty$ and $\mathbf{1}_{\{\tau_a < \tau_b\}} = 0$ a.s. In effect, this removes the lower exit level, and the corresponding expected discounted reward is

$$\mathbb{E}_x\{e^{-r(\tau_a \wedge \tau_b)}h(X_{\tau_a \wedge \tau_b})\}$$
$$= \mathbb{E}_x\{e^{-r\tau_a}h(X_{\tau_a})\mathbf{1}_{\{\tau_a < \tau_b\}}\} + \mathbb{E}_x\{e^{-r\tau_b}h(X_{\tau_b})\mathbf{1}_{\{\tau_a > \tau_b\}}\}$$
$$= \mathbb{E}_x\{e^{-r\tau_b}h(X_{\tau_b})\}.$$

Consequently, by considering interval-type strategies, we also include the class of stopping strategies of reaching a single upper level b (see Theorem 2.6 below).

Next, we prove the optimality of the proposed stopping strategy and provide an expression for the value function.

Theorem 2.3. *The value function $V(x)$ defined in (2.3) is given by*

$$V(x) = G(x)W(\psi(x)), \tag{2.19}$$

where G, ψ and W are defined in (2.12), (2.13) and (2.18), respectively.

Proof. Since $\tau_a \wedge \tau_b \in \mathcal{T}$, we have

$$V(x) \geq \sup_{\{a,b:a \leq x \leq b\}} \mathbb{E}_x\{e^{-r(\tau_a \wedge \tau_b)}h(X_{\tau_a \wedge \tau_b})\} = G(x)W(\psi(x)).$$

To show the reverse inequality, we first show that

$$G(x)W(\psi(x)) \geq \mathbb{E}_x\{e^{-r(t \wedge \tau)}G(X_{t \wedge \tau})W(\psi(X_{t \wedge \tau}))\},$$

for $\tau \in \mathcal{T}$ and $t \geq 0$. The concavity of W implies that, for any fixed z, there exists an affine function $L_z(\alpha) := m_z \alpha + c_z$ such that $L_z(\alpha) \geq W(\alpha)$ and $L_z(z) = W(z)$ at $\alpha = z$, where m_z and c_z are both constants depending on z. This leads to the inequality

$$\mathbb{E}_x\{e^{-r(t \wedge \tau)}G(X_{t \wedge \tau})W(\psi(X_{t \wedge \tau}))\}$$
$$\leq \mathbb{E}_x\{e^{-r(t \wedge \tau)}G(X_{t \wedge \tau})L_{\psi(x)}(\psi(X_{t \wedge \tau}))\}$$
$$= m_{\psi(x)}\mathbb{E}_x\{e^{-r(t \wedge \tau)}G(X_{t \wedge \tau})\psi(X_{t \wedge \tau})\} + c_{\psi(x)}\mathbb{E}_x\{e^{-r(t \wedge \tau)}G(X_{t \wedge \tau})\}$$
$$= m_{\psi(x)}\mathbb{E}_x\{e^{-r(t \wedge \tau)}F(X_{t \wedge \tau})\} + c_{\psi(x)}\mathbb{E}_x\{e^{-r(t \wedge \tau)}G(X_{t \wedge \tau})\}$$
$$= m_{\psi(x)}F(x) + c_{\psi(x)}G(x) \tag{2.20}$$
$$= G(x)L_{\psi(x)}(\psi(x))$$
$$= G(x)W(\psi(x)), \tag{2.21}$$

where (2.20) follows from the martingale property of $(e^{-rt}F(X_t))_{t \geq 0}$ and $(e^{-rt}G(X_t))_{t \geq 0}$.

By (2.21) and the fact that W majorizes H, it follows that

$$G(x)W(\psi(x)) \geq \mathbb{E}_x\{e^{-r(t\wedge\tau)}G(X_{t\wedge\tau})W(\psi(X_{t\wedge\tau}))\}$$
$$\geq \mathbb{E}_x\{e^{-r(t\wedge\tau)}G(X_{t\wedge\tau})H(\psi(X_{t\wedge\tau}))\}$$
$$= \mathbb{E}_x\{e^{-r(t\wedge\tau)}h(X_{t\wedge\tau})\}. \tag{2.22}$$

Maximizing (2.22) over all $\tau \in \mathcal{T}$ and $t \geq 0$ yields that $G(x)W(\psi(x)) \geq V(x)$. $\qquad\square$

Let us emphasize that the optimal levels (a^*, b^*) may depend on the initial value x, and can potentially coincide, or take values $-\infty$ and $+\infty$. As such, the structure of the stopping and delay regions can potentially be characterized by multiple intervals, leading to *disconnected* delay regions (see Theorem 2.17 below).

We follow the procedure for Theorem 2.3 to derive the expression for the value function J in (2.4). First, we denote $\hat{F}(x) = F(x; \hat{r})$ and $\hat{G}(x) = G(x; \hat{r})$ (see (2.11)–(2.12)), with discount rate $\hat{r}$. In addition, we define the transformation

$$\hat{\psi}(x) := \frac{\hat{F}}{\hat{G}}(x) \quad \text{and} \quad \hat{h}(x) = V(x) - x - c_b. \tag{2.23}$$

Using these functions, we consider the function analogous to H:

$$\hat{H}(z) := \begin{cases} \dfrac{\hat{h}}{\hat{G}} \circ \hat{\psi}^{-1}(z) & \text{if } z > 0, \\[2mm] \lim\limits_{x \to -\infty} \dfrac{(\hat{h}(x))^+}{\hat{G}(x)} & \text{if } z = 0. \end{cases} \tag{2.24}$$

Following the steps (2.14)–(2.18) with F, G, ψ, and H replaced by $\hat{F}$, $\hat{G}$, $\hat{\psi}$, and $\hat{H}$, respectively, we write down the smallest concave majorant $\hat{W}$ of $\hat{H}$, namely,

$$\hat{W}(z) := \sup_{\{z_{\hat{a}}, z_{\hat{b}} : z_{\hat{a}} \leq z \leq z_{\hat{b}}\}} \left[\hat{H}(z_{\hat{a}}) \frac{z_{\hat{b}} - z}{z_{\hat{b}} - z_{\hat{a}}} + \hat{H}(z_{\hat{b}}) \frac{z - z_{\hat{a}}}{z_{\hat{b}} - z_{\hat{a}}} \right].$$

From this, we seek to determine the candidate optimal entry interval $(z_{\hat{a}^*}, z_{\hat{b}^*})$ in the $z = \hat{\psi}(x)$ coordinate. Following the proof of Theorem 2.3 with the new functions $\hat{F}$, $\hat{G}$, $\hat{\psi}$, $\hat{H}$, and $\hat{W}$, the value function of the optimal entry timing problem admits the expression

$$J(x) = \hat{G}(x)\hat{W}(\hat{\psi}(x)). \tag{2.25}$$

An alternative way to solve for $V(x)$ and $J(x)$ is to look for the solutions to the pair of variational inequalities

$$\min\{rV(x) - \mathcal{L}V(x), V(x) - (x - c_s)\} = 0, \tag{2.26}$$
$$\min\{\hat{r}J(x) - \mathcal{L}J(x), J(x) - (V(x) - x - c_b)\} = 0, \tag{2.27}$$

for $x \in \mathbb{R}$. With sufficient regularity conditions, this approach can verify that the solutions to the VIs, $V(x)$ and $J(x)$, indeed correspond to the optimal stopping problems (see, for example, Theorem 10.4.1 of Øksendal (2003)). Nevertheless, this approach does not immediately suggest candidate optimal timing strategies or value functions, and typically begins with a conjecture on the structure of the optimal stopping times, followed by verification. In contrast, our approach allows us to directly construct the value functions, at the cost of analyzing the properties of H, W, $\hat{H}$, and $\hat{W}$.

2.4 Analytical Results

We will first study the optimal exit timing in Section 2.4.1, followed by the optimal entry timing problem in Section 2.4.2.

2.4.1 *Optimal Exit Timing*

We now analyze the optimal exit timing problem (2.3) under the OU model. First, we obtain a bound for the value function V in terms of F.

Lemma 2.4. *There exists a positive constant K such that, for all $x \in \mathbb{R}$, $0 \leq V(x) \leq KF(x)$.*

In preparation for the next result, we summarize the crucial properties of H.

Lemma 2.5. *The function H is continuous on $[0, +\infty)$, twice differentiable on $(0, +\infty)$ and possesses the following properties:*

(i) $H(0) = 0$, and

$$H(z) \begin{cases} < 0 & \text{if } z \in (0, \psi(c_s)), \\ > 0 & \text{if } z \in (\psi(c_s), +\infty). \end{cases}$$

(ii) Let x^* be the unique solution to $G(x) - (x - c_s)G'(x) = 0$. Then, we have

$$H(z) \text{ is strictly} \begin{cases} \text{decreasing} & \text{if } z \in (0, \psi(x^*)), \\ \text{increasing} & \text{if } z \in (\psi(x^*), +\infty), \end{cases}$$

and $x^* < c_s \wedge L^*$ with

$$L^* = \frac{\mu\theta + rc_s}{\mu + r}. \tag{2.28}$$

(iii)

$$H(z) \text{ is } \begin{cases} convex & if \ z \in (0, \psi(L^*)], \\ concave & if \ z \in [\psi(L^*), +\infty). \end{cases}$$

Based on Lemma 2.5, we sketch H in Figure 2.2. The properties of H are essential in deriving the value function and optimal liquidation level, as we show next.

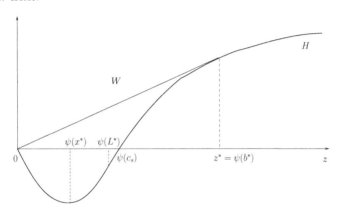

Fig. 2.2 Sketches of H and W. By Lemma 2.5, H is convex on the left of $\psi(L^*)$ and concave on the right. The smallest concave majorant W is a straight line tangent to H at z^* on $[0, z^*)$, and coincides with H on $[z^*, +\infty)$.

Theorem 2.6. *The optimal liquidation problem (2.3) admits the solution*

$$V(x) = \begin{cases} (b^* - c_s)\frac{F(x)}{F(b^*)} & if \ x \in (-\infty, b^*), \\ x - c_s & otherwise, \end{cases} \qquad (2.29)$$

where the optimal liquidation level b^* is found from the equation

$$F(b) = (b - c_s)F'(b), \qquad (2.30)$$

and is bounded below by $L^* \vee c_s$. The corresponding optimal liquidation time is given by

$$\tau^* = \inf\{t \geq 0 : X_t \geq b^*\}. \qquad (2.31)$$

Proof. From Lemma 2.5 and the fact that $H'(z) \to 0$ as $z \to +\infty$ (see also Figure 2.2), we infer that there exists a unique number $z^* > \psi(L^*) \vee \psi(c_s)$ such that

$$\frac{H(z^*)}{z^*} = H'(z^*). \qquad (2.32)$$

In turn, the smallest concave majorant is given by

$$W(z) = \begin{cases} z\frac{H(z^*)}{z^*} & \text{if } z < z^*, \\ H(z) & \text{if } z \geq z^*. \end{cases} \qquad (2.33)$$

Substituting $b^* = \psi^{-1}(z^*)$ into (2.32), we have the LHS

$$\frac{H(z^*)}{z^*} = \frac{H(\psi(b^*))}{\psi(b^*)} = \frac{b^* - c_s}{F(b^*)}, \qquad (2.34)$$

and the RHS

$$\begin{aligned} H'(z^*) &= \frac{G(\psi^{-1}(z^*)) - (\psi^{-1}(z^*) - c_s)G'(\psi^{-1}(z^*))}{F'(\psi^{-1}(z^*))G(\psi^{-1}(z^*)) - F(\psi^{-1}(z^*))G'(\psi^{-1}(z^*))} \\ &= \frac{G(b^*) - (b^* - c_s)G'(b^*)}{F'(b^*)G(b^*) - F(b^*)G'(b^*)}. \end{aligned}$$

Equivalently, we can express condition (2.32) in terms of b^*:

$$\frac{b^* - c_s}{F(b^*)} = \frac{G(b^*) - (b^* - c_s)G'(b^*)}{F'(b^*)G(b^*) - F(b^*)G'(b^*)},$$

which can be further simplified to

$$F(b^*) = (b^* - c_s)F'(b^*).$$

Applying (2.34) to (2.33), we get

$$W(\psi(x)) = \begin{cases} \psi(x)\frac{H(z^*)}{z^*} = \frac{F(x)}{G(x)}\frac{b^* - c_s}{F(b^*)} & \text{if } x < b^*, \\ H(\psi(x)) = \frac{x - c_s}{G(x)} & \text{if } x \geq b^*. \end{cases} \qquad (2.35)$$

In turn, we obtain the value function $V(x)$ by substituting (2.35) into (2.19). $\qquad \square$

Next, we examine the dependence of the investor's optimal timing strategy on the transaction cost c_s.

Proposition 2.7. *The value function $V(x)$ of (2.3) is decreasing in the transaction cost c_s for every $x \in \mathbb{R}$, and the optimal liquidation level b^* is increasing in c_s.*

Proof. For any $x \in \mathbb{R}$ and $\tau \in \mathcal{T}$, the corresponding expected discounted reward, $\mathbb{E}_x\{e^{-r\tau}(X_\tau - c_s)\} = \mathbb{E}_x\{e^{-r\tau}X_\tau\} - c_s\,\mathbb{E}_x\{e^{-r\tau}\}$, is decreasing in c_s. This implies that $V(x)$ is also decreasing in c_s. Next, we treat the optimal threshold $b^*(c_s)$ as a function of c_s, and differentiate (2.30) w.r.t. c_s to get

$$b^{*\prime}(c_s) = \frac{F'(b^*)}{(b^* - c_s)F''(b^*)} > 0.$$

Since $F'(x) > 0$, $F''(x) > 0$ (see (2.11)), and $b^* > c_s$ according to Theorem 2.6, we conclude that b^* is increasing in c_s. $\qquad \square$

In other words, if the transaction cost is high, the investor would tend to liquidate at a higher level, in order to compensate for the loss on transaction cost.

For other parameters, such as μ and σ, the dependence of b^* is illustrated numerically. In Figure 2.3(a), the optimal exit level b^* is plotted against the speed of mean reversion μ with different long-run means θ. First, we observe that the optimal exit level increases with the long-run mean. We also note that b^* is decreasing with the speed of mean reversion. This means if the price returns to the mean faster, then the investor should liquidate at a lower level.

In Figure 2.3(b), the optimal exit level b^* is plotted against the volatility parameter σ with different long-run means. It confirms that the optimal exit level increases with the long-run mean. In addition, we observe that the optimal exit level increases as volatility increases. The more volatile the process is, the more probable it is to reach a level further away from the mean, which gives us an opportunity to make a profit from liquidating from a higher level.

2.4.2 *Optimal Entry Timing*

Having solved for the optimal exit timing, we now turn to the optimal entry timing problem. In this case, the value function is

$$J(x) = \sup_{\nu \in \mathcal{T}} \mathbb{E}_x\{e^{-\hat{r}\nu}(V(X_\nu) - X_\nu - c_b)\}, \quad x \in \mathbb{R},$$

where $V(x)$ is given by Theorem 2.6.

Lemma 2.8. *There exists a positive constant $\hat{K}$ such that, for all $x \in \mathbb{R}$, $0 \le J(x) \le \hat{K}\hat{G}(x)$.*

To solve for the optimal entry threshold(s), we will need several properties of $\hat{H}$, as we summarize below.

Lemma 2.9. *The function $\hat{H}$ is continuous on $[0, +\infty)$, differentiable on $(0, +\infty)$, and twice differentiable on $(0, \hat{\psi}(b^*)) \cup (\hat{\psi}(b^*), +\infty)$, and possesses the following properties:*

(i) $\hat{H}(0) = 0$. Let $\bar{d}$ denote the unique solution to $\hat{h}(x) = 0$, then $\bar{d} < b^*$ and

$$\hat{H}(z) \begin{cases} > 0 & \text{if } z \in (0, \hat{\psi}(\bar{d})), \\ < 0 & \text{if } z \in (\hat{\psi}(\bar{d}), +\infty). \end{cases}$$

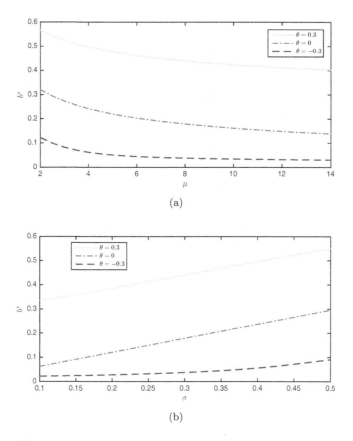

Fig. 2.3   (a) The optimal liquidation level b^* vs speed of mean reversion μ. Parameters: $\sigma = 0.3$, $r = 0.05$, $c_s = 0.02$. (b) The optimal liquidation level b^* vs volatility σ. Parameters: $\mu = 8$, $r = 0.05$, $c_s = 0.02$.

(ii) $\hat{H}(z)$ is strictly decreasing if $z \in (\hat{\psi}(b^*), +\infty)$.

(iii) Let $\underline{b}$ denote the unique solution to $(\mathcal{L} - \hat{r})\hat{h}(x) = 0$, then $\underline{b} < L^*$ and

$$
\hat{H}(z) \text{ is} \begin{cases} concave & \textit{if } z \in (0, \hat{\psi}(\underline{b})), \\ convex & \textit{if } z \in (\hat{\psi}(\underline{b}), +\infty). \end{cases}
$$

In Figure 2.4, we give a sketch of $\hat{H}$ according to Lemma 2.9. This will be useful for deriving the optimal entry level.

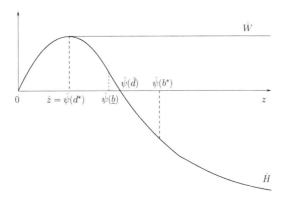

Fig. 2.4 Sketches of $\hat{H}$ and $\hat{W}$. The function $\hat{W}$ coincides with $\hat{H}$ on $[0, \hat{z}]$ and is equal to the constant $\hat{H}(\hat{z})$ on $(\hat{z}, +\infty)$.

Theorem 2.10. *The optimal entry timing problem* (2.4) *admits the solution*

$$J(x) = \begin{cases} V(x) - x - c_b & \text{if } x \in (-\infty, d^*], \\ \frac{V(d^*) - d^* - c_b}{\hat{G}(d^*)} \hat{G}(x) & \text{if } x \in (d^*, +\infty), \end{cases} \tag{2.36}$$

where the optimal entry level d^* *is found from the equation*

$$\hat{G}(d)(V'(d) - 1) = \hat{G}'(d)(V(d) - d - c_b). \tag{2.37}$$

Proof. We look for the value function of the form: $J(x) = \hat{G}(x)\hat{W}(\hat{\psi}(x))$, where $\hat{W}$ is the the smallest concave majorant of $\hat{H}$. From Lemma 2.9 and Figure 2.4, we infer that there exists a unique number $\hat{z} < \hat{\psi}(b^*)$ such that

$$\hat{H}'(\hat{z}) = 0. \tag{2.38}$$

This implies that

$$\hat{W}(z) = \begin{cases} \hat{H}(z) & \text{if } z \leq \hat{z}, \\ \hat{H}(\hat{z}) & \text{if } z > \hat{z}. \end{cases} \tag{2.39}$$

Substituting $d^* = \hat{\psi}^{-1}(\hat{z})$ into (2.38), we have

$$\hat{H}'(\hat{z}) = \frac{\hat{G}(d^*)(V'(d^*) - 1) - \hat{G}'(d^*)(V(d^*) - d^* - c_b)}{\hat{F}'(d^*)\hat{G}(d^*) - \hat{F}(d^*)\hat{G}'(d^*)} = 0,$$

which is equivalent to condition (2.37). Furthermore, using (2.23) and (2.24), we get

$$\hat{H}(\hat{z}) = \frac{V(d^*) - d^* - c_b}{\hat{G}(d^*)}. \tag{2.40}$$

To conclude, we substitute $\hat{H}(\hat{z})$ of (2.40) and $\hat{H}(z)$ of (2.24) into $\hat{W}$ of (2.39), which by (2.25) yields the value function $J(x)$ in (2.36). $\square$

With the analytic solutions for V and J, we can verify by direct substitution that $V(x)$ in (2.29) and $J(x)$ in (2.36) satisfy both (2.26) and (2.27).

Since the optimal entry timing problem is nested with another optimal stopping problem, the parameter dependence of the optimal entry level is complicated. Below, we illustrate the impact of transaction cost.

Proposition 2.11. *The optimal entry level d^* of* (2.4) *is decreasing in the transaction cost c_b.*

Proof. Considering the optimal entry level d^* as a function of c_b, we differentiate (2.37) w.r.t. c_b to get

$$d^{*\prime}(c_b) = \frac{-\hat{G}'(d^*)}{\hat{G}(d^*)}[V''(d^*) - \frac{V(d^*) - d^* - c_b}{\hat{G}(d^*)}\hat{G}''(d^*)]^{-1}. \qquad (2.41)$$

Since $\hat{G}(d^*) > 0$ and $\hat{G}'(d^*) < 0$, the sign of $d^{*\prime}(c_b)$ is determined by $V''(d^*) - \frac{V(d^*)-d^*-c_b}{\hat{G}(d^*)}\hat{G}''(d^*)$. Denote $\hat{f}(x) = \frac{V(d^*)-d^*-c_b}{\hat{G}(d^*)}\hat{G}(x)$. Recall that $\hat{h}(x) = V(x) - x - c_b$,

$$J(x) = \begin{cases} \hat{h}(x) & \text{if } x \in (-\infty, d^*], \\ \hat{f}(x) > \hat{h}(x) & \text{if } x \in (d^*, +\infty), \end{cases}$$

and $\hat{f}(x)$ smooth pastes $\hat{h}(x)$ at d^*. Since both $\hat{h}(x)$ and $\hat{f}(x)$ are positive decreasing convex functions, it follows that $\hat{h}''(d^*) \leq \hat{f}''(d^*)$. Observing that $\hat{h}''(d^*) = V''(d^*)$ and $\hat{f}''(d^*) = \frac{V(d^*)-d^*-c_b}{\hat{G}(d^*)}\hat{G}''(d^*)$, we have $V''(d^*) - \frac{V(d^*)-d^*-c_b}{\hat{G}(d^*)}\hat{G}''(d^*) \leq 0$. Applying this to (2.41), we conclude that $d^{*\prime}(c_b) \leq 0$. $\qquad \square$

We numerically examine the dependence of d^* on other parameters in Figure 2.5. These optimal entry levels are computed with the same set of parameters as those for the optimal exit level in Figure 2.3. Naturally, d^* increases with the long-run mean θ. The dependence of d^* on μ and σ is exactly the opposite from that of b^*. For better illustration, we plot both the optimal entry level d^* and the optimal exit level b^* against μ and σ in Figure 2.6. We see that the faster the speed of mean reversion, the closer the buy and sell levels. On the other hand, the higher the volatility, the further apart the buy and sell levels. As we have discussed earlier, the more likely the process reaches levels further away from the mean, the wider the gap between buy and sell levels. In other words, when the volatility is high,

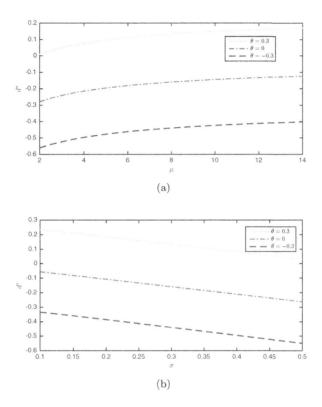

(a)

(b)

Fig. 2.5 (a) The optimal entry level d^* vs speed of mean reversion μ. Parameters: $\sigma = 0.3$, $r = \hat{r} = 0.05$, $c_s = c_b = 0.02$. (b) The optimal entry level d^* vs volatility σ. Parameters: $\mu = 8$, $r = \hat{r} = 0.05$, $c_s = c_b = 0.02$.

it is possible for the investor to delay both entry and exit times to seek a wider spread.

Remark 2.12. We end this section with a special example in the OU model with $\mu = 0$ in (2.1). It follows that X reduces to a Brownian motion: $X_t = \sigma B_t$, $t \geq 0$. In this case, the optimal liquidation level b^* for problem (2.3) is

$$b^* = c_s + \frac{\sigma}{\sqrt{2r}},$$

and the optimal entry level d^* for problem (2.4) is the root to the equation

$$\left(1 + \sqrt{\frac{\hat{r}}{r}}\right) e^{\frac{\sqrt{2r}}{\sigma}(d - c_s - \frac{\sigma}{\sqrt{2r}})} = \frac{\sqrt{2\hat{r}}}{\sigma}(d + c_b) + 1, \quad d \in (-\infty, b^*).$$

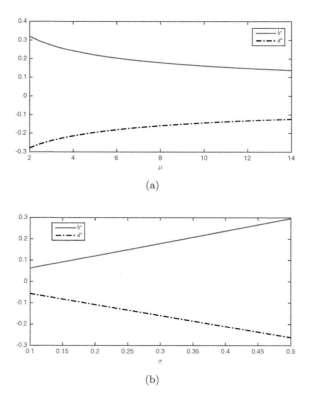

Fig. 2.6 (a) The optimal entry level d^* and the optimal liquidation level b^* vs speed of mean reversion μ. Parameters: $\theta = 0$, $\sigma = 0.3$, $r = \hat{r} = 0.05$, $c_s = c_b = 0.02$. (b) The optimal entry level d^* and the optimal liquidation level b^* vs volatility σ. Parameters: $\theta = 0$, $\mu = 8$, $r = \hat{r} = 0.05$, $c_s = c_b = 0.02$.

2.5 Incorporating Stop-Loss Exit

Now we consider the optimal entry and exit problems with a stop-loss constraint. For convenience, we restate the value functions from (2.5) and (2.6):

$$J_L(x) = \sup_{\nu \in \mathcal{T}} \mathbb{E}_x \left\{ e^{-\hat{r}\nu} (V_L(X_\nu) - X_\nu - c_b) \right\}, \qquad (2.42)$$

$$V_L(x) = \sup_{\tau \in \mathcal{T}} \mathbb{E}_x \left\{ e^{-r(\tau \wedge \tau_L)} (X_{\tau \wedge \tau_L} - c_s) \right\}. \qquad (2.43)$$

After solving for the optimal timing strategies, we will also examine the dependence of the optimal liquidation threshold on the stop-loss level L.

2.5.1 *Optimal Exit Timing*

We first give an analytic solution to the optimal exit timing problem.

Theorem 2.13. *The optimal liquidation problem* (2.43) *with stop-loss level* L *admits the solution*

$$V_L(x) = \begin{cases} CF(x) + DG(x) & \text{if } x \in (L, b_L^*), \\ x - c_s & \text{otherwise,} \end{cases} \tag{2.44}$$

where

$$C = \frac{(b_L^* - c_s)G(L) - (L - c_s)G(b_L^*)}{F(b_L^*)G(L) - F(L)G(b_L^*)}, \quad D = \frac{(L - c_s)F(b_L^*) - (b_L^* - c_s)F(L)}{F(b_L^*)G(L) - F(L)G(b_L^*)}.$$

The optimal liquidation level b_L^* *is found from the equation*

$$[(L - c_s)G(b) - (b - c_s)G(L)]F'(b) + [(b - c_s)F(L) - (L - c_s)F(b)]G'(b)$$
$$= G(b)F(L) - G(L)F(b). \tag{2.45}$$

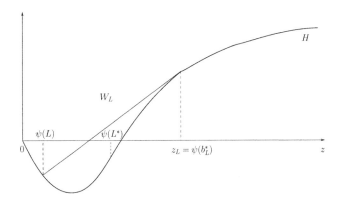

Fig. 2.7 Sketch of W_L. On $[0, \psi(L)] \cup [z_L, +\infty)$, W_L coincides with H, and over $(\psi(L), z_L)$, W_L is a straight line tangent to H at z_L.

Proof. Due to the stop-loss level L, we consider the smallest concave majorant of $H(z)$, denoted by $W_L(z)$, over the restricted domain $[\psi(L), +\infty)$ and set $W_L(z) = H(z)$ for $z \in [0, \psi(L)]$.

From Lemma 2.5 and Figure 2.7, we see that $H(z)$ is convex over $(0, \psi(L^*)]$ and concave in $[\psi(L^*), +\infty)$. If $L \geq L^*$, then $H(z)$ is concave over $[\psi(L), +\infty)$, which implies that $W_L(z) = H(z)$ for $z \geq 0$, and thus

$V_L(x) = x - c_s$ for $x \in \mathbb{R}$. On the other hand, if $L < L^*$, then $H(z)$ is convex on $[\psi(L), \psi(L^*)]$, and concave strictly increasing on $[\psi(L^*), +\infty)$. There exists a unique number $z_L > \psi(L^*)$ such that

$$\frac{H(z_L) - H(\psi(L))}{z_L - \psi(L)} = H'(z_L). \qquad (2.46)$$

In turn, the smallest concave majorant admits the form:

$$W_L(z) = \begin{cases} H(\psi(L)) + (z - \psi(L))H'(z_L) & \text{if } z \in (\psi(L), z_L), \\ H(z) & \text{otherwise.} \end{cases} \qquad (2.47)$$

Substituting $b_L^* = \psi^{-1}(z_L)$ into (2.46), we have from the LHS

$$\frac{H(z_L) - H(\psi(L))}{z_L - \psi(L)} = \frac{H(\psi(b_L^*)) - H(\psi(L))}{\psi(b_L^*) - \psi(L)} = \frac{\frac{b_L^* - c_s}{G(b_L^*)} - \frac{L - c_s}{G(L)}}{\frac{F(b_L^*)}{G(b_L^*)} - \frac{F(L)}{G(L)}} = C,$$

and the RHS

$$\begin{aligned} H'(z_L) &= \frac{G(\psi^{-1}(z_L)) - (\psi^{-1}(z_L) - c_s)G'(\psi^{-1}(z_L))}{F'(\psi^{-1}(z_L))G(\psi^{-1}(z_L)) - F(\psi^{-1}(z_L))G'(\psi^{-1}(z_L))} \\ &= \frac{G(b_L^*) - (b^* - c_s)G'(b_L^*)}{F'(b_L^*)G(b_L^*) - F(b_L^*)G'(b_L^*)}. \end{aligned}$$

Therefore, we can equivalently express (2.46) in terms of b_L^*:

$$\frac{(b_L^* - c_s)G(L) - (L - c_s)G(b_L^*)}{F(b_L^*)G(L) - F(L)G(b_L^*)} = \frac{G(b_L^*) - (b_L^* - c_s)G'(b_L^*)}{F'(b_L^*)G(b_L^*) - F(b_L^*)G'(b_L^*)},$$

which by rearrangement immediately simplifies to (2.45).

Furthermore, for $x \in (L, b_L^*)$, $H'(z_L) = C$ implies that

$$W_L(\psi(x)) = H(\psi(L)) + (\psi(x) - \psi(L))C.$$

Substituting this to $V_L(x) = G(x)W_L(\psi(x))$, the value function becomes

$$\begin{aligned} V_L(x) &= G(x)\big[H(\psi(L)) + (\psi(x) - \psi(L))C\big] \\ &= CF(x) + G(x)\big[H(\psi(L)) - \psi(L)C\big], \end{aligned}$$

which resembles (2.44) after the observation that

$$\begin{aligned} H(\psi(L)) - \psi(L)C &= \frac{L - c_s}{G(L)} - \frac{F(L)}{G(L)}\frac{(b_L^* - c_s)G(L) - (L - c_s)G(b_L^*)}{F(b_L^*)G(L) - F(L)G(b_L^*)} \\ &= \frac{(L - c_s)F(b_L^*) - (b_L^* - c_s)F(L)}{F(b_L^*)G(L) - F(L)G(b_L^*)} = D. \end{aligned}$$

$\square$

We can interpret the investor's timing strategy in terms of three price intervals, namely, the liquidation region $[b_L^*, +\infty)$, the delay region (L, b_L^*), and the stop-loss region $(-\infty, L]$. In both liquidation and stop-loss regions, the value function $V_L(x) = x - c_s$, and therefore, the investor will immediately close out the position. From the proof of Theorem 2.13, if $L \geq L^* = \frac{\mu\theta + rc_s}{\mu + r}$ (see (2.28)), then $V_L(x) = x - c_s$, $\forall x \in \mathbb{R}$. In other words, if the stop-loss level is too high, then the delay region completely disappears, and the investor will liquidate immediately for every initial value $x \in \mathbb{R}$.

Corollary 2.14. *If $L < L^*$, then there exists a unique solution $b_L^* \in (L^*, +\infty)$ that solves (2.45). If $L \geq L^*$, then $V_L(x) = x - c_s$, for $x \in \mathbb{R}$.*

The direct effect of a stop-loss exit constraint is forced liquidation whenever the price process reaches L before the upper liquidation level b_L^*. Interestingly, there is an additional indirect effect: a higher stop-loss level will induce the investor to *voluntarily* liquidate earlier at a lower take-profit level.

Proposition 2.15. *The optimal liquidation level b_L^* of (2.43) strictly decreases as the stop-loss level L increases.*

Proof. Recall that $z_L = \psi(b_L^*)$ and ψ is a strictly increasing function. Therefore, it is sufficient to show that z_L strictly decreases as $\tilde{L} := \psi(L)$ increases. As such, we denote $z_L(\tilde{L})$ to highlight its dependence on $\tilde{L}$. Differentiating (2.46) w.r.t. $\tilde{L}$ gives

$$z_L'(\tilde{L}) = \frac{H'(z_L) - H'(\tilde{L})}{H''(z_L)(z_L - \tilde{L})}. \tag{2.48}$$

It follows from the definitions of W_L and z_L that $H'(z_L) > H'(\tilde{L})$ and $z_L > \tilde{L}$. Also, we have $H''(z) < 0$ since H is concave at z_L. Applying these to (2.48), we conclude that $z_L'(\tilde{L}) < 0$. $\qquad\square$

Figure 2.8 illustrates the optimal exit price level b_L^* as a function of the stop-loss levels L, for different long-run means θ. When b_L^* is strictly greater than L (on the left of the straight line), the delay region is nonempty. As L increases, b_L^* strictly decreases and the two meet at L^* (on the straight line), and the delay region vanishes.

Also, there is an interesting connection between cases with different long-run means and transaction costs. To this end, let us denote the value

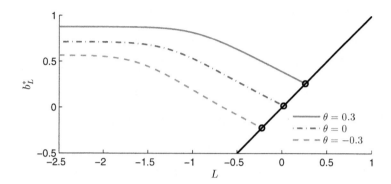

Fig. 2.8   The optimal exit threshold b_L^* is strictly decreasing with respect to the stop-loss level L. The straight line is where $b_L^* = L$, and each of the three circles locates the critical stop-loss level L^*.

function by $V_L(x; \theta, c_s)$ to highlight the dependence on θ and c_s, and the corresponding optimal liquidation level by $b_L^*(\theta, c_s)$. We find that, for any $\theta_1, \theta_2 \in \mathbb{R}$, $c_1, c_2 > 0$, $L_1 \le \frac{\mu\theta_1 + rc_1}{\mu + r}$, and $L_2 \le \frac{\mu\theta_2 + rc_2}{\mu + r}$, the associated value functions and optimal liquidation levels satisfy the relationships:

$$V_{L_1}(x + \theta_1; \theta_1, c_1) = V_{L_2}(x + \theta_2; \theta_2, c_2), \qquad (2.49)$$

$$b_{L_1}^*(\theta_1, c_1) - \theta_1 = b_{L_2}^*(\theta_2, c_2) - \theta_2, \qquad (2.50)$$

as long as $\theta_1 - \theta_2 = c_1 - c_2 = L_1 - L_2$. These results (2.49) and (2.50) also hold in the case without stop-loss.

2.5.2   Optimal Entry Timing

We now discuss the optimal entry timing problem $J_L(x)$ defined in (2.42). Since $\sup_{x \in \mathbb{R}}(V_L(x) - x - c_b) \le 0$ implies that $J_L(x) = 0$ for $x \in \mathbb{R}$, we can focus on the case with

$$\sup_{x \in \mathbb{R}}(V_L(x) - x - c_b) > 0, \qquad (2.51)$$

and look for non-trivial optimal timing strategies.

Associated with reward function $\hat{h}_L(x) := V_L(x) - x - c_b$ from entering the market, we define the function $\hat{H}_L$ as in (2.17) whose properties are summarized in the following lemma.

Lemma 2.16. *The function $\hat{H}_L$ is continuous on $[0, +\infty)$, differentiable on $(0, \hat{\psi}(L)) \cup (\hat{\psi}(L), +\infty)$, twice differentiable on $(0, \hat{\psi}(L)) \cup (\hat{\psi}(L), \hat{\psi}(b_L^*)) \cup (\hat{\psi}(b_L^*), +\infty)$, and possesses the following properties:*

(i) $\hat{H}_L(0) = 0$. $\hat{H}_L(z) < 0$ for $z \in (0, \hat{\psi}(L)] \cup [\hat{\psi}(b_L^*), +\infty)$.

(ii) $\hat{H}_L(z)$ is strictly decreasing for $z \in (0, \hat{\psi}(L)) \cup (\hat{\psi}(b_L^*), +\infty)$.

(iii) There exists some constant $\bar{d}_L \in (L, b_L^*)$ such that $(\mathcal{L} - \hat{r})\hat{h}_L(\bar{d}_L) = 0$, and

$$\hat{H}_L(z) \text{ is } \begin{cases} convex & if \ z \in (0, \hat{\psi}(L)) \cup (\hat{\psi}(\bar{d}_L), +\infty), \\ concave & if \ z \in (\hat{\psi}(L), \hat{\psi}(\bar{d}_L)). \end{cases}$$

In addition, $\hat{z}_1 \in (\hat{\psi}(L), \hat{\psi}(\bar{d}_L))$, where $\hat{z}_1 := \arg\max_{z \in [0, +\infty)} \hat{H}_L(z)$.

Theorem 2.17. *The optimal entry timing problem (2.42) admits the solution*

$$J_L(x) = \begin{cases} P\hat{F}(x) & if \ x \in (-\infty, a_L^*), \\ V_L(x) - x - c_b & if \ x \in [a_L^*, d_L^*], \\ Q\hat{G}(x) & if \ x \in (d_L^*, +\infty), \end{cases} \tag{2.52}$$

where

$$P = \frac{V_L(a_L^*) - a_L^* - c_b}{\hat{F}(a_L^*)}, \quad Q = \frac{V_L(d_L^*) - d_L^* - c_b}{\hat{G}(d_L^*)}.$$

The optimal entry time is given by

$$\nu_{a_L^*, d_L^*} = \inf\{t \geq 0 : X_t \in [a_L^*, d_L^*]\}, \tag{2.53}$$

where the critical levels a_L^* and d_L^* satisfy, respectively,

$$\hat{F}(a)(V_L'(a) - 1) = \hat{F}'(a)(V_L(a) - a - c_b), \tag{2.54}$$

and

$$\hat{G}(d)(V_L'(d) - 1) = \hat{G}'(d)(V_L(d) - d - c_b). \tag{2.55}$$

Proof. We look for the value function of the form: $J_L(x) = \hat{G}(x)\hat{W}_L(\hat{\psi}(x))$, where $\hat{W}_L$ is the smallest non-negative concave majorant of $\hat{H}_L$. From Lemma 2.16 and the sketch of $\hat{H}_L$ in Figure 2.9, the maximizer of $\hat{H}_L$, $\hat{z}_1$, satisfies

$$\hat{H}_L'(\hat{z}_1) = 0. \tag{2.56}$$

Also there exists a unique number $\hat{z}_0 \in (\hat{\psi}(L), \hat{z}_1)$ such that

$$\frac{\hat{H}_L(\hat{z}_0)}{\hat{z}_0} = \hat{H}_L'(\hat{z}_0). \tag{2.57}$$

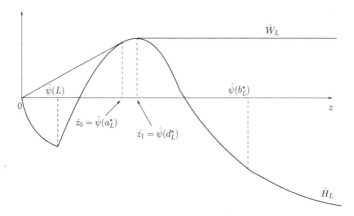

Fig. 2.9 Sketches of $\hat{H}_L$ and $\hat{W}_L$. $\hat{W}_L$ is a straight line tangent to $\hat{H}_L$ at $\hat{z}_0$ on $[0, \hat{z}_0)$, coincides with $\hat{H}_L$ on $[\hat{z}_0, \hat{z}_1]$, and is equal to the constant $\hat{H}_L(\hat{z}_1)$ on $(\hat{z}_1, +\infty)$. Note that $\hat{H}_L$ is not differentiable at $\hat{\psi}(L)$.

In turn, the smallest non-negative concave majorant admits the form:

$$\hat{W}_L(z) = \begin{cases} z\hat{H}'_L(\hat{z}_0) & \text{if } z \in [0, \hat{z}_0), \\ \hat{H}_L(z) & \text{if } z \in [\hat{z}_0, \hat{z}_1], \\ \hat{H}_L(\hat{z}_1) & \text{if } z \in (\hat{z}_1, +\infty). \end{cases}$$

Substituting $a_L^* = \hat{\psi}^{-1}(\hat{z}_0)$ into (2.57), we have

$$\frac{\hat{H}_L(\hat{z}_0)}{\hat{z}_0} = \frac{V_L(a_L^*) - a_L^* - c_b}{\hat{F}(a_L^*)},$$

$$\hat{H}'_L(\hat{z}_0) = \frac{\hat{G}(a_L^*)(V'_L(a_L^*) - 1) - \hat{G}'(a_L^*)(V_L(a_L^*) - a_L^* - c_b)}{\hat{F}'(a_L^*)\hat{G}(a_L^*) - \hat{F}(a_L^*)\hat{G}'(a_L^*)}.$$

Equivalently, we can express condition (2.57) in terms of a_L^*:

$$\frac{V_L(a_L^*) - a_L^* - c_b}{\hat{F}(a_L^*)} = \frac{\hat{G}(a_L^*)(V'_L(a_L^*) - 1) - \hat{G}'(a_L^*)(V_L(a_L^*) - a_L^* - c_b)}{\hat{F}'(a_L^*)\hat{G}(a_L^*) - \hat{F}(a_L^*)\hat{G}'(a_L^*)}.$$

Simplifying this shows that a_L^* solves (2.54). Also, we can express $\hat{H}'_L(\hat{z}_0)$ in terms of a_L^*:

$$\hat{H}'_L(\hat{z}_0) = \frac{\hat{H}_L(\hat{z}_0)}{\hat{z}_0} = \frac{V_L(a_L^*) - a_L^* - c_b}{\hat{F}(a_L^*)} = P.$$

In addition, substituting $d_L^* = \hat{\psi}^{-1}(\hat{z}_1)$ into (2.56), we have

$$\hat{H}'_L(\hat{z}_1) = \frac{\hat{G}(d_L^*)(V'_L(d_L^*) - 1) - \hat{G}'(d_L^*)(V_L(d_L^*) - d_L^* - c_b)}{\hat{F}'(d_L^*)\hat{G}(d_L^*) - \hat{F}(d_L^*)\hat{G}'(d_L^*)} = 0,$$

which, after a straightforward simplification, is identical to (2.55).

Also, $\hat{H}_L(\hat{z}_1)$ can be written as

$$\hat{H}_L(\hat{z}_1) = \frac{V_L(d_L^*) - d_L^* - c_b}{\hat{G}(d_L^*)} = Q.$$

Substituting these to $J_L(x) = \hat{G}(x)\hat{W}_L(\hat{\psi}(x))$, we arrive at (2.52). $\qquad\square$

Theorem 2.17 reveals that the optimal entry region is characterized by a price interval $[a_L^*, d_L^*]$ strictly above the stop-loss level L and strictly below the optimal exit level b_L^*. In particular, if the current asset price is between L and a_L^*, then it is optimal for the investor to wait even though the price is low. This is intuitive because if the entry price is too close to L, then the investor is very likely to be forced to exit at a loss afterwards. Moreover, delaying to enter also discounts the transaction cost. As a consequence, the investor's delay region, where she would wait to enter the market, is disconnected.

Figure 2.10 illustrates two simulated paths and the associated exercise times. We have chosen L to be 2 standard deviations below the long-run mean θ, with other parameters from our pairs trading example. By Theorem 2.17, the investor will enter the market at $\nu_{a_L^*, d_L^*}$ (see (2.53)). Since both paths start with $X_0 > d_L^*$, the investor waits to enter until the OU path reaches d_L^* from above, as indicated by ν_d^* in panels (a) and (b). After entry, Figure 2.10(a) describes the scenario where the investor exits voluntarily at the optimal level b_L^*, whereas in Figure 2.10(b) the investor is forced to exit at the stop-loss level L. These optimal levels are calculated from Theorems 2.13 and 2.17 based on the given estimated parameters.

Lastly, we remark that the optimal levels a_L^*, d_L^* and b_L^* are outputs of the models, depending on the parameters (μ, θ, σ) and the choice of stop-loss level L. Recall that our model parameters are estimated based on the likelihood maximizing portfolio discussed in Section 2.1. Other estimation methodologies and price data can be used, and may lead to different portfolio strategies (α, β) and estimated parameters values (μ, θ, σ). In turn, the resulting optimal entry and exit thresholds may also change accordingly.

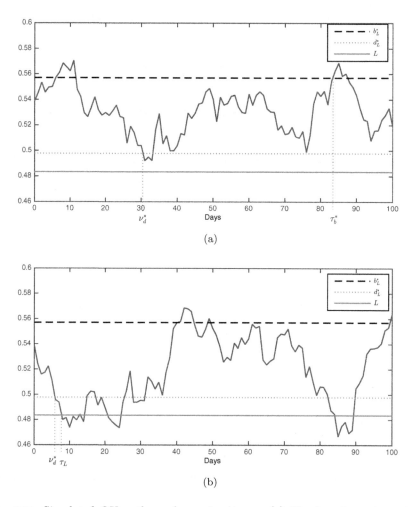

Fig. 2.10 Simulated OU paths and exercise times. (a) The investor enters at $\nu_d^* = \inf\{t \geq 0 : X_t \leq d_L^*\}$ with $d_L^* = 0.4978$, and exit at $\tau_b^* = \inf\{t \geq \nu_d^* : X_t \geq b_L^*\}$ with $b_L^* = 0.5570$. (b) The investor enters at $\nu_d^* = \inf\{t \geq 0 : X_t \leq d_L^*\}$ but exits at stop-loss level $L = 0.4834$. Parameters: $\theta = 0.5388$, $\mu = 16.6677$, $\sigma = 0.1599$, $r = \hat{r} = 0.05$, and $c_s = c_b = 0.05$.

2.5.3   *Relative Stop-Loss Exit*

For some investors, it may be more desirable to set the stop-loss contingent on the entry level. In other words, if the value of X at the entry time is x, then the investor would assign a lower stop-loss level $x - \ell$, for some constant $\ell > 0$. Therefore, the investor faces the optimal entry timing problem

$$\mathcal{J}_\ell(x) = \sup_{\nu \in \mathcal{T}} \mathbb{E}_x \left\{ e^{-\hat{r}\nu} (\mathcal{V}_\ell(X_\nu) - X_\nu - c_b) \right\},$$

where $\mathcal{V}_\ell(x) := V_{x-\ell}(x)$ (see (2.43)) is the optimal exit timing problem with stop-loss level $x - \ell$. The dependence of $V_{x-\ell}(x)$ on x is significantly more complicated than $V(x)$ or $V_L(x)$, making the problem much less tractable.

In Figure 2.11, we illustrate numerically the optimal timing strategies. The investor will still enter at a lower level d^*. After entry, the investor will wait to exit at either the stop-loss level $d^* - \ell$ or an upper level b^*.

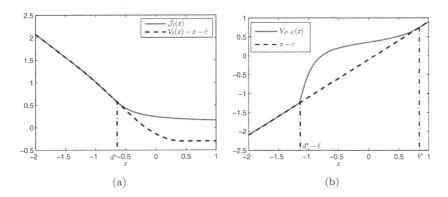

(a)   (b)

Fig. 2.11   (a) The optimal entry value function $\mathcal{J}_\ell(x)$ dominates the reward function $\mathcal{V}_\ell(x) - x - c_b$, and they coincide for $x \leq d^*$. (b) For the exit problem, the stop-loss level is $d^* - \ell$ and the optimal liquidation level is b^*.

2.5.4   *Optimal Switching with Stop-Loss Exit*

As an alternative to our double stopping approach to trading under mean reversion, the optimal switching approach assumes that the investor commits to an infinite number of trades prior to reaching the stop-loss level. This problem has been studied by Song and Zhang (2013), and we summarize their main result here.

The sequential trading times are modeled by the stopping times $\nu_1, \tau_1, \nu_2, \tau_2, \cdots \in \mathcal{T}$ such that $0 \leq \nu_1 \leq \tau_1 \leq \nu_2 \leq \tau_2 \leq \cdots \leq \tau_L$. A share of the risky asset is bought and sold, respectively, at times ν_i and τ_i, $i \in \mathbb{N}$. The investor's optimal timing to trade would depend on the initial position. Precisely, if the investor starts with a zero position, then the first trading decision is when to *buy* and the optimal switching problem is

$$\tilde{J}_L(x) = \sup_{\Lambda_0} \mathbb{E}_x \left\{ \sum_{n=1}^{\infty} [e^{-r\tau_n}(X_{\tau_n} - c) - e^{-r\nu_n}(X_{\nu_n} + c)]\mathbf{1}_{\{\nu_n < \tau_L\}} \right\},$$

with the set of admissible stopping times $\Lambda_0 = (\nu_1, \tau_1, \nu_2, \tau_2, \dots)$. As in Song and Zhang (2013), the same transactions cost, denoted by c, is incurred for buying and selling the asset.

On the other hand, if the investor is initially holding a share of the asset, then the investor first determines when to *sell*, and subsequently switch between buy and sell positions. This leads to the optimal switching problem:

$$\tilde{V}_L(x) = \sup_{\Lambda_1} \mathbb{E}_x \left\{ e^{-r\tau_1}(X_{\tau_1} - c) \right.$$
$$\left. + \sum_{n=2}^{\infty} [e^{-r\tau_n}(X_{\tau_n} - c) - e^{-r\nu_n}(X_{\nu_n} + c)]\mathbf{1}_{\{\nu_n < \tau_L\}} \right\},$$

with $\Lambda_1 = (\tau_1, \nu_2, \tau_2, \nu_3, \dots)$. Note that the first stopping time τ_1 is the time to sell the asset for the first time, and (ν_i, τ_i) are the stopping times, respectively, for buying and selling for the ith time.

Theorem 2.18. *Song and Zhang (2013) Let* $\tilde{a}_L^*, \tilde{d}_L^*, \tilde{b}_L^*, \tilde{A}, \tilde{B}_1, \tilde{B}_2, \tilde{C}_1, \tilde{C}_2$ *be solutions of the non-linear system of equations:*

$$\tilde{B}_1 F(\tilde{a}_L^*) + \tilde{B}_2 G(\tilde{a}_L^*) = \tilde{C}_1 F(\tilde{a}_L^*) + \tilde{C}_2 G(\tilde{a}_L^*) - \tilde{a}_L^* - c,$$
$$\tilde{B}_1 F'(\tilde{a}_L^*) + \tilde{B}_2 G'(\tilde{a}_L^*) = \tilde{C}_1 F'(\tilde{a}_L^*) + \tilde{C}_2 G'(\tilde{a}_L^*) - 1,$$
$$\tilde{A} G(\tilde{d}_L^*) = \tilde{C}_1 F(\tilde{d}_L^*) + \tilde{C}_2 G(\tilde{d}_L^*) - \tilde{d}_L^* - c,$$
$$\tilde{A} G'(\tilde{d}_L^*) = \tilde{C}_1 F'(\tilde{d}_L^*) + \tilde{C}_2 G'(\tilde{d}_L^*) - 1,$$
$$\tilde{C}_1 F(\tilde{b}_L^*) + \tilde{C}_2 G(\tilde{b}_L^*) = \tilde{A} G(\tilde{b}_L^*) + \tilde{b}_L^* - c,$$
$$\tilde{C}_1 F'(\tilde{b}_L^*) + \tilde{C}_2 G'(\tilde{b}_L^*) = \tilde{A} G'(\tilde{b}_L^*) + 1,$$
$$\tilde{B}_1 F(L) + \tilde{B}_2 G(L) = 0,$$
$$\tilde{C}_1 F(L) + \tilde{C}_2 G(L) = L - c,$$

and satisfy

$$\tilde{d}_L^* \le \frac{\mu\theta - rc}{\mu + r}, \quad and \quad \tilde{b}_L^* \ge L^* = \frac{\mu\theta + rc}{\mu + r},$$

and

$$|(\tilde{C}_1 - \tilde{B}_1 F(x) + (\tilde{C}_2 - \tilde{B}_2 G(x) - x| \le c \quad on \ (L, \tilde{a}_L^*),$$
$$|\tilde{C}_1 F(x) + (\tilde{C}_2 - \tilde{A})G(x) - x| \le c \quad on \ (\tilde{d}_L^*, \tilde{b}_L^*).$$

In addition, let

$$\tilde{j}_L = \begin{cases} \tilde{B}_1 F(x) + \tilde{B}_2 G(x) & if \ x \in (L, \tilde{a}_L^*), \\ \tilde{C}_1 F(x) + \tilde{C}_2 G(x) - x - c & if \ x \in [\tilde{a}_L^*, \tilde{d}_L^*], \\ \tilde{A}G(x) & if \ x \in (\tilde{d}_L^*, +\infty), \end{cases}$$

$$\tilde{v}_L = \begin{cases} \tilde{C}_1 F(x) + \tilde{C}_2 G(x) & if \ x \in (L, \tilde{d}_L^*), \\ \tilde{A}G(x) + x - c & if \ x \in (\tilde{b}_L^*, +\infty), \end{cases}$$

and assume $\tilde{j}_L \ge 0$. Then, $\tilde{j}_L(x) = \tilde{J}_L(x)$, and $\tilde{v}_L(x) = \tilde{V}_L(x)$.

Moreover, (i) if the investor starts with a zero position, let $\Lambda_0^* = (\nu_1^*, \tau_1^*, \nu_2^*, \tau_2^*, \dots)$, such that the stopping times, for $i \ge 1$,

$$\nu_1^* = \inf\{t \ge 0 : X_t \in [\tilde{a}_L^*, \tilde{d}_L^*]\} \wedge \tau_L,$$

$$\tau_i^* = \inf\{t > \nu_i^* : X_t \notin (L, \tilde{b}_L^*)\} \wedge \tau_L,$$

$$\nu_{i+1}^* = \inf\{t > \tau_i^* : X_t \in [\tilde{a}_L^*, \tilde{d}_L^*]\} \wedge \tau_L.$$

(ii) If the investor starts with a long position, let $\Lambda_1^* = (\tau_1^*, \nu_2^*, \tau_2^*, \nu_3^*, \dots)$, such that, $i \ge 2$,

$$\tau_1^* = \inf\{t \ge 0 : X_t \notin (L, \tilde{b}_L^*)\} \wedge \tau_L,$$

$$\nu_i^* = \inf\{t > \tau_{i-1}^* : X_t \in [\tilde{a}_L^*, \tilde{d}_L^*]\} \wedge \tau_L,$$

$$\tau_i^* = \inf\{t > \nu_i^* : X_t \notin (L, \tilde{b}_L^*)\} \wedge \tau_L.$$

Then Λ_0^* and Λ_1^* are optimal to cases (i) and (ii), respectively.

2.6 Further Applications

The optimal trading problem studied herein is amenable for a number of extensions. Our model can be considered as the building block for the problem with any finite number of sequential trades. The major challenge lies in analyzing and computing the value functions for optimal entry and exit, as we have done for the first ones, namely, J and V. We will conclude by briefly discussing the incorporation of a minimum holding period or a timing penalty.

2.6.1   *Minimum Holding Period*

Another timing constraint of practical interest is the minimum holding period. Recently, regulators and exchanges are contemplating to apply this rule to rein in high-frequency trading. This gives rise to the need to better understand the effect of this restriction on trading. Intuitively, a minimum holding period always delays the liquidation timing, but how does it influence the investor's timing to *enter* the market?

Suppose that once the investor enters the position, she is only allowed to liquidate after a pre-specified time period δ. The incorporation of a minimum holding period leads to the constrained optimal stopping problem

$$V^\delta(x) = \sup_{\tau \geq \delta} \mathbb{E}_x\{e^{-r\tau}(X_\tau - c_s)\} = \mathbb{E}_x\{e^{-r\delta}V(X_\delta)\},$$

where $V(x)$ the unconstrained problem in (2.3) with solution given in Theorem 2.6. The second equality follows from the strong Markov property of X and the optimality of $V(x)$. Compared to the unconstrained problem, the optimal liquidation timing for $V^\delta(x)$ is simply delayed by δ but otherwise identical to τ^* in (2.31). Also, by the supermartingale and non-negative property of $V(x)$, we see that $0 \leq V^\delta(x) \leq V(x)$ and $V^\delta(x)$ decreases with δ.

Turning to the optimal entry timing, the investor solves

$$J^\delta(x) = \sup_{\nu \in \mathcal{T}} \mathbb{E}_x\{e^{-\hat{r}\nu}(V^\delta(X_\nu) - X_\nu - c_b)\}. \tag{2.58}$$

The following result reflects the impact of the minimum holding period.

Proposition 2.19. *For every $x \in \mathbb{R}$, we have $J^\delta(x) \leq J(x)$ and $d^\delta \leq d^*$.*

This means that the minimum holding period leads to a lower optimal entry level and lower value function as compared to the original value function J in (2.4). Next, we present the proof.

Proof. As in Theorem 2.10, one can show that the optimal entry timing problem (2.58) admits the solution

$$J^\delta(x) = \begin{cases} V^\delta(x) - x - c_b & \text{if } x \in (-\infty, d^\delta], \\ \frac{V^\delta(d^\delta) - d^\delta - c_b}{\hat{G}(d^\delta)}\hat{G}(x) & \text{if } x \in (d^\delta, +\infty), \end{cases}$$

where the optimal entry level d^δ is found from the equation

$$\hat{G}(d)(V^{\delta'}(d) - 1) = \hat{G}'(d)(V^\delta(d) - d - c_b).$$

To compare with the original case, we first define $h_2(x) = -x - c_b$,

$$\hat{H}^\delta(z) = (\frac{V^\delta + h_2}{\hat{G}}) \circ \hat{\psi}^{-1}(z),$$

and denote $\hat{W}^\delta(z)$ as the smallest concave majorant of $\hat{H}^\delta(z)$. Following the similar proof of Theorem 2.10, we can show that

$$\hat{W}^\delta(z) = \begin{cases} \hat{H}^\delta(z) & \text{if } z \in [0, \hat{z}^\delta], \\ \hat{H}^\delta(\hat{z}^\delta) & \text{if } z \in (\hat{z}^\delta, +\infty), \end{cases}$$

where $\hat{z}^\delta = \hat{\psi}(d^\delta)$ satisfies $\hat{H}^{\delta'}(\hat{z}^\delta) = 0$. Recall that $\hat{z} = \hat{\psi}(d^*)$ satisfies $\hat{H}'(\hat{z}) = 0$.

To show $d^\delta \leq d^*$, we examine the concavity of $\hat{H}^\delta$ and $\hat{H}$. Restating $\hat{H}$ in (2.24) in terms of h_2:

$$\hat{H}(z) = (\frac{V + h_2}{\hat{G}}) \circ \hat{\psi}^{-1}(z),$$

followed by differentiation, we have

$$\hat{H}''(z) = \frac{2}{\sigma^2 \hat{G}(x)(\hat{\psi}'(x))^2} (\mathcal{L} - \hat{r})(V + h_2)(x), \quad z = \hat{\psi}(x). \quad (2.59)$$

Similarly, (2.59) also holds for $\hat{H}^\delta$ with V replaced by V^δ. This leads us to analyze $(\mathcal{L} - \hat{r})(V + h_2)(x)$ and $(\mathcal{L} - \hat{r})(V^\delta + h_2)(x)$. As shown in Lemma 2.9 and Figure 2.4, $\hat{H}(z)$ is concave for $z \in (0, \hat{\psi}(\underline{b}))$, where $\underline{b} < L^*$ satisfies $(\mathcal{L} - \hat{r})(V + h_2)(x) = 0$, and $\hat{z} < \hat{\psi}(\underline{b})$.

Moreover, it follows from the supermartingale property of V that

$$\mathbb{E}_x\{e^{-rt}V^\delta(X_t)\} = \mathbb{E}_x\{e^{-r(t+\delta)}V(X_{t+\delta})\} \leq \mathbb{E}_x\{e^{-r\delta}V(X_\delta)\} = V^\delta(x).$$

From this and Proposition 5.9 in Dayanik and Karatzas (2003), we infer that $(\mathcal{L} - r)V^\delta(x) \leq 0$. In turn, for $x < \underline{b}$, we have

$$\begin{aligned} (\mathcal{L} - \hat{r})(V^\delta + h_2)(x) &= (\mathcal{L} - r)V^\delta(x) + (r - \hat{r})V^\delta(x) + (\mathcal{L} - \hat{r})h_2(x) \\ &\leq (r - \hat{r})V^\delta(x) + (\mathcal{L} - \hat{r})h_2(x) \\ &\leq (r - \hat{r})V(x) + (\mathcal{L} - \hat{r})h_2(x) \\ &= (\mathcal{L} - \hat{r})(V + h_2)(x), \end{aligned}$$

where the last equality follows from the fact that $(\mathcal{L} - r)V(x) = 0$ for $x < b^*$, since W is a straight line for $z \leq \psi(b^*)$, and $\underline{b} < L^* < b^*$. Hence, for $z \in (0, \hat{\psi}(\underline{b}))$, $\hat{H}^{\delta''}(z) \leq \hat{H}''(z) \leq 0$ and $\hat{H}^\delta(z)$ is also concave.

Since $V(x) \geq V^\delta(x) \geq 0$, we have $\hat{H}(z) \geq \hat{H}^\delta(z)$ for $z \in (0, +\infty)$. Considering $\hat{H}(0) = \hat{H}^\delta(0) = 0$ and $\hat{H}(0+), \hat{H}^\delta(0+) > 0$, we have $\hat{H}'(0+) \geq$

$\hat{H}^{\delta'}(0+) \geq 0$. This, along with $\hat{H}^{\delta''}(z) \leq \hat{H}''(z) \leq 0$ for $z \in (0, \hat{\psi}(\underline{b}))$, imply that $\hat{H}'(z) \geq \hat{H}^{\delta'}(z)$ for $z \in (0, \hat{\psi}(\underline{b}))$. So $\hat{H}^{\delta'}(\hat{z}) \leq \hat{H}'(\hat{z}) = 0$. Considering $\hat{H}^{\delta'}(\hat{z}^\delta) = 0$ and the concavity of $\hat{H}^\delta$, we conclude that $\hat{z}^\delta \leq \hat{z}$, which by the monotonicity of $\hat{\psi}$ is equivalent to $d^\delta \leq d^*$.

To show $J^\delta(x) \leq J(x)$, it is equivalent to establish $\hat{W}^\delta(z) \leq \hat{W}(z)$ for all $z \in [0, \infty)$: (i) For $z \in [0, \hat{z}^\delta]$, this holds since $\hat{W}^\delta(z) = \hat{H}^\delta(z)$, $\hat{W}(z) = \hat{H}(z)$, and $\hat{H}^\delta(z) \leq \hat{H}(z)$. (ii) For $z \in (\hat{z}^\delta, \hat{z}]$, $\hat{W}^\delta(z) = \hat{H}^\delta(\hat{z}^\delta) \leq \hat{H}(\hat{z}^\delta) \leq \hat{H}(z) = \hat{W}(z)$, where the last inequality follows from the fact that $\hat{H}'(z) \geq 0$ for $z \in (\hat{z}^\delta, \hat{z}]$. (iii) For $z \in (\hat{z}, +\infty)$, $\hat{W}^\delta(z) = \hat{H}^\delta(\hat{z}^\delta) \leq \hat{H}(\hat{z}) = \hat{W}(z)$. $\square$

2.6.2 *Path-Dependent Risk Penalty*

In addition to maximizing the expected liquidation value, a risk-sensitive investor may be concerned about the price fluctuation over time, and therefore, be willing to adjust her liquidation timing depending on the path behavior of prices. This motivates the incorporation of a *path-dependent* risk penalty up to the liquidation time τ. To illustrate this idea, we apply a penalty term of the form $\mathbb{E}_x\{\int_0^\tau e^{-ru} q(X_u)\, du\}$, where $q(x)$ could be any positive penalty function. This risk penalty only applies when the investor is holding the position, but not before entry.

Hence, the investor solves the penalized optimal timing problems:

$$\mathcal{J}^q(x) = \sup_{\nu \in \mathcal{T}} \mathbb{E}_x\left\{ e^{-r\nu}(V^q(X_\nu) - X_\nu - c_b) \right\},$$

$$V^q(x) = \sup_{\tau \in \mathcal{T}} \mathbb{E}_x\left\{ e^{-r\tau}(X_\tau - c_s) - \int_0^\tau e^{-ru} q(X_u)\, du \right\}. \qquad (2.60)$$

As a special case, let $q(x) \equiv q$, a strictly positive constant. Then, by computing the integral in (2.60),

$$V^q(x) = \sup_{\tau \in \mathcal{T}} \mathbb{E}_x\left\{ e^{-r\tau}(X_\tau - (c_s - \frac{q}{r})) \right\} - \frac{q}{r}. \qquad (2.61)$$

This presents an interesting connection between the penalized problem $V^q(x)$ in (2.61) and the unpenalized optimal stopping problem V in (2.3). Indeed, we observe that the penalty term amounts to reducing the transaction cost c_s by the positive constant $\frac{q}{r}$. In other words, the optimal stopping time τ_q^* for $V^q(x)$ is identical to the optimal stopping time τ^* for $V(x)$ in (2.3) but with c_s replaced by $c_s - \frac{q}{r}$. Furthermore, since b^* is increasing in c_s, a higher penalty q lowers the optimal liquidation level. As for the entry problem $\mathcal{J}^q$, the solution is found from Theorem 2.10 by modifying the transaction cost to be $c_b + \frac{q}{r}$. More sophisticated path-dependent risk penalties can be considered under this formulation, including those based on the (integrated) shortfall with $q(x) = \rho((m-x)^+)$ where m is a constant benchmark and ρ is an increasing convex loss function (see, for example, Section 4.9 of Föllmer and Schied (2004)).

2.7 Proofs of Lemmas

In this last section, we present the proofs to a number of lemmas and propositions regarding the properties of V, J, H and $\hat{H}$.

Proof of Lemma 2.4 (Bounds of V)

First, observe that $F(-\infty) = G(+\infty) = 0$ and $F(+\infty) = G(-\infty) = +\infty$. The limit

$$\limsup_{x \to +\infty} \frac{(h(x))^+}{F(x)} = \limsup_{x \to +\infty} \frac{x - c_s}{F(x)} = \limsup_{x \to +\infty} \frac{1}{F'(x)} = 0.$$

Therefore, there exists some x_0 such that $(h(x))^+ < F(x)$ for $x \in (x_0, +\infty)$. As for $x \leq x_0$, $(h(x))^+$ is bounded above by the constant $(x_0 - c_s)^+$. As a result, we can always find a constant K such that $(h(x))^+ \leq KF(x)$ for all $x \in \mathbb{R}$.

By definition, the process $(e^{-rt}F(X_t))_{t \geq 0}$ is a martingale. This implies, for every $x \in \mathbb{R}$ and $\tau \in \mathcal{T}$,

$$KF(x) = \mathbb{E}_x\{e^{-r\tau}KF(X_\tau)\} \geq \mathbb{E}_x\{e^{-r\tau}(h(X_\tau))^+\} \geq \mathbb{E}_x\{e^{-r\tau}h(X_\tau)\}.$$

Therefore, $V(x) \leq KF(x)$. Lastly, the choice of $\tau = +\infty$ as a candidate stopping time implies that $V(x) \geq 0$.

Proof of Lemma 2.5 (Properties of H)

The continuity and twice differentiability of H on $(0, +\infty)$ follow directly from those of h, G and ψ. To show the continuity of H at 0, since $H(0) = \lim_{x \to -\infty} \frac{(x-c_s)^+}{G(x)} = 0$, we only need to show that $\lim_{y \to 0} H(z) = 0$. Note that $z = \psi(x) \to 0$, as $x \to -\infty$. Therefore,

$$\lim_{z \to 0} H(z) = \lim_{x \to -\infty} \frac{h(x)}{G(x)} = \lim_{x \to -\infty} \frac{x - c_s}{G(x)} = \lim_{x \to -\infty} \frac{1}{G'(x)} = 0.$$

We conclude that H is also continuous at 0.

(i) One can show that $\psi(x) \in (0, +\infty)$ for $x \in \mathbb{R}$ and is a strictly increasing function. Then property (i) follows directly from the fact that $G(x) > 0$.

(ii) By the definition of H,

$$H'(z) = \frac{1}{\psi'(x)}\left(\frac{h}{G}\right)'(x) = \frac{h'(x)G(x) - h(x)G'(x)}{\psi'(x)G^2(x)}, \quad z = \psi(x).$$

Since both $\psi'(x)$ and $G^2(x)$ are positive, we only need to determine the sign of $h'(x)G(x) - h(x)G'(x) = G(x) - (x - c_s)G'(x)$.

Define $u(x) := (x - c_s) - \frac{G(x)}{G'(x)}$. Note that $u(x) + c_s$ is the intersecting point at x axis of the tangent line of $G(x)$, and $u'(x) = \frac{G(x)G''(x)}{(G'(x))^2}$. Since $G(\cdot)$ is a positive, strictly decreasing and convex function, $u(x)$ is strictly increasing and $u(x) < 0$ as $x \to -\infty$. Also, note that

$$u(c_s) = -\frac{G(c_s)}{G'(c_s)} > 0,$$

$$u(L^*) = (L^* - c_s) - \frac{G(x)}{G'(x)} = \frac{\mu}{r}(\theta - L^*) - \frac{G(L^*)}{G'(L^*)} = -\frac{\sigma^2}{2r}\frac{G''(L^*)}{G'(L^*)} > 0.$$

Therefore, there exists a unique root x^* that solves $u(x) = 0$, and $x^* < c_s \wedge L^*$, such that

$$G(x) - (x - c_s)G'(x) \begin{cases} < 0 & \text{if } x \in (-\infty, x^*), \\ > 0 & \text{if } x \in (x^*, +\infty). \end{cases}$$

Thus $H(z)$ is strictly decreasing if $z \in (0, \psi(x^*))$, and increasing otherwise.

(iii) By the definition of H,

$$H''(z) = \frac{2}{\sigma^2 G(x)(\psi'(x))^2}(\mathcal{L} - r)h(x), \quad z = \psi(x).$$

Since $\sigma^2, G(x)$ and $(\psi'(x))^2$ are all positive, we only need to determine the sign of $(\mathcal{L} - r)h(x)$:

$$(\mathcal{L} - r)h(x) = \mu(\theta - x) - r(x - c_s)$$

$$= (\mu\theta + rc_s) - (\mu + r)x \begin{cases} \geq 0 & \text{if } x \in (-\infty, L^*], \\ \leq 0 & \text{if } x \in [L^*, +\infty). \end{cases}$$

Therefore, $H(z)$ is convex if $z \in (0, \psi(L^*)]$, and concave otherwise.

Proof of Lemma 2.8 (Bounds of J)

Since $\hat{F}(-\infty) = \hat{G}(+\infty) = 0$ and $\hat{F}(+\infty) = \hat{G}(-\infty) = +\infty$. Next, from the limit

$$\limsup_{x \to -\infty} \frac{\left(\hat{h}(x)\right)^+}{\hat{G}(x)} = \limsup_{x \to -\infty} \frac{\hat{h}(x)}{\hat{G}(x)} = 0,$$

we see that there exists some $\hat{x}_0$ such that $(\hat{h}(x))^+ < \hat{G}(x)$ for every $x \in (-\infty, \hat{x}_0)$. Since $(\hat{h}(x))^+$ is bounded between $[0, (V(\hat{x}_0) - \hat{x}_0 - c_b)^+]$ for $x \in [\hat{x}_0, +\infty)$, there exists some constant $\hat{K}$ such that $(\hat{h}(x))^+ \leq \hat{K}\hat{G}(x)$ for all $x \in \mathbb{R}$.

By the definition of $\hat{G}$, we can write $\hat{G}(x) = \mathbb{E}_x\{e^{-\hat{r}\tau}\hat{G}(X_\tau)\}$ for any $\tau \in \mathcal{T}$. This yields the inequality

$$\hat{K}\hat{G}(x) = \mathbb{E}_x\{e^{-\hat{r}\tau}\hat{K}\hat{G}(X_\tau)\} \geq \mathbb{E}_x\{e^{-\hat{r}\tau}(\hat{h}(X_\tau))^+\} \geq \mathbb{E}_x\{e^{-\hat{r}\tau}\hat{h}(X_\tau)\},$$

for every $x \in \mathbb{R}$ and every $\tau \in \mathcal{T}$. Hence, $J(x) \leq \hat{K}\hat{G}(x)$. Since $\tau = +\infty$ is a candidate stopping time, we have $J(x) \geq 0$.

Proof of Lemma 2.9 (Properties of $\hat{H}$)

We first show that $V(x)$ and $\hat{h}(x)$ are twice differentiable everywhere, except for $x = b^*$. Recall that

$$V(x) = \begin{cases} (b^* - c_s)\frac{F(x)}{F(b^*)} & \text{if } x \in (-\infty, b^*), \\ x - c_s & \text{otherwise,} \end{cases} \quad \text{and} \quad \hat{h}(x) = V(x) - x - c_b.$$

Therefore, it follows from (2.30) that

$$V'(x) = \begin{cases} (b^* - c_s)\frac{F'(x)}{F(b^*)} = \frac{F'(x)}{F'(b^*)} & \text{if } x \in (-\infty, b^*), \\ 1 & \text{if } x \in (b^*, +\infty), \end{cases}$$

which implies that $V'(b^*-) = 1 = V'(b^*+)$. Therefore, $V(x)$ is differentiable everywhere and so is $\hat{h}$. However, $V(x)$ is not twice differentiable since

$$V''(x) = \begin{cases} \frac{F''(x)}{F'(b^*)} & \text{if } x \in (-\infty, b^*), \\ 0 & \text{if } x \in (b^*, +\infty), \end{cases}$$

and $V''(b^*-) \neq V''(b^*+)$. Consequently, $\hat{h}(x) = V(x) - x - c_b$ is not twice differentiable at b^*.

The twice differentiability of $\hat{G}$ and $\hat{\psi}$ are straightforward. The continuity and differentiability of $\hat{H}$ on $(0, +\infty)$ and twice differentiability on $(0, \hat{\psi}(b^*)) \cup (\hat{\psi}(b^*), +\infty)$ follow directly. Observing that $\hat{h}(x) > 0$ as $x \to -\infty$, $\hat{H}$ is also continuous at 0 by definition. We now establish the properties of $\hat{H}$.

(i) First we prove the value of $\hat{H}$ at 0:

$$\hat{H}(0) = \lim_{x \to -\infty} \frac{(\hat{h}(x))^+}{\hat{G}(x)} = \limsup_{x \to -\infty} \frac{\frac{(b^* - c_s)}{F(b^*)}F(x) - x - c_b}{\hat{G}(x)}$$

$$= \limsup_{x \to -\infty} \frac{\frac{(b^* - c_s)}{F(b^*)}F'(x) - 1}{\hat{G}'(x)} = 0.$$

Next, observe that $\lim_{x\to-\infty} \hat{h}(x) = +\infty$ and $\hat{h}(x) = -(c_s + c_b)$, for $x \in [b^*, +\infty)$. Since $F'(x)$ is strictly increasing and $F'(x) > 0$ for $x \in \mathbb{R}$, we have, for $x < b^*$,

$$\hat{h}'(x) = V'(x) - 1 = \frac{F'(x)}{F'(b^*)} - 1 < \frac{F'(b^*)}{F'(b^*)} - 1 = 0,$$

which implies that $\hat{h}(x)$ is strictly decreasing for $x \in (-\infty, b^*)$. Therefore, there exists a unique solution $\bar{d}$ to $\hat{h}(x) = 0$, and $\bar{d} < b^*$, such that $\hat{h}(x) > 0$ if $x \in (-\infty, \bar{d})$ and $\hat{h}(x) < 0$ if $x \in (\bar{d}, +\infty)$. It is trivial that $\hat{\psi}(x) \in (0, +\infty)$ for $x \in \mathbb{R}$ and is a strictly increasing function. Therefore, along with the fact that $\hat{G}(x) > 0$, property (i) follows directly.

(ii) With $z = \hat{\psi}(x)$, for $x > b^*$,

$$\hat{H}'(z) = \frac{1}{\hat{\psi}'(x)}(\frac{\hat{h}}{\hat{G}})'(x) = \frac{1}{\hat{\psi}'(x)}(\frac{-(c_s + c_b)}{\hat{G}(x)})' = \frac{1}{\hat{\psi}'(x)}\frac{(c_s + c_b)\hat{G}'(x)}{\hat{G}^2(x)} < 0,$$

since $\hat{\psi}'(x) > 0$, $\hat{G}'(x) < 0$, and $\hat{G}^2(x) > 0$. Therefore, $\hat{H}(z)$ is strictly decreasing for $z > \hat{\psi}(b^*)$.

(iii) By the definition of $\hat{H}$,

$$\hat{H}''(z) = \frac{2}{\sigma^2 \hat{G}(x)(\hat{\psi}'(x))^2}(\mathcal{L} - \hat{r})\hat{h}(x), \quad z = \hat{\psi}(x).$$

Since σ^2, $\hat{G}(x)$ and $(\hat{\psi}'(x))^2$ are all positive, we only need to determine the sign of $(\mathcal{L} - \hat{r})\hat{h}(x)$:

$$(\mathcal{L} - \hat{r})\hat{h}(x) = \frac{1}{2}\sigma^2 V''(x) + \mu(\theta - x)V'(x) - \mu(\theta - x) - \hat{r}(V(x) - x - c_b)$$

$$= \begin{cases} (r - \hat{r})V(x) + (\mu + \hat{r})x - \mu\theta + \hat{r}c_b & \text{if } x < b^*, \\ \hat{r}(c_s + c_b) > 0 & \text{if } x > b^*. \end{cases}$$

To determine the sign of $(\mathcal{L} - \hat{r})\hat{h}(x)$ in $(-\infty, b^*)$, first note that $[(\mathcal{L} - \hat{r})\hat{h}](x)$ is a strictly increasing function in $(-\infty, b^*)$, since $V(x)$ is a strictly increasing function and $r \geq \hat{r}$ by assumption. Next note that for $x \in [L^*, b^*)$,

$$(\mathcal{L} - \hat{r})\hat{h}(x) = (r - \hat{r})V(x) + (\mu + \hat{r})x - \mu\theta + \hat{r}c_b$$
$$\geq (r - \hat{r})(x - c_s) + (\mu + \hat{r})x - \mu\theta + \hat{r}c_b$$
$$= (r + \mu)x - (\mu\theta + rc_s) + \hat{r}(c_s + c_b)$$
$$\geq (r + \mu)L^* - (\mu\theta + rc_s) + \hat{r}(c_s + c_b) = \hat{r}(c_s + c_b) > 0.$$

Also, note that $(\mathcal{L} - \hat{r})\hat{h}(x) \to -\infty$ as $x \to -\infty$. Therefore, $(\mathcal{L} - \hat{r})\hat{h}(x) < 0$ if $x \in (-\infty, \underline{b})$ and $(\mathcal{L} - \hat{r})\hat{h}(x) > 0$ if $x \in (\underline{b}, +\infty)$ with $\underline{b} < L^*$ being the break-even point. From this, we conclude property (iii).

Proof of Lemma 2.16 (Properties of $\hat{H}_L$)

(i) The continuity of $\hat{H}_L(z)$ on $(0, +\infty)$ is implied by the continuities of $\hat{h}_L$, $\hat{G}$ and $\hat{\psi}$. The continuity of $\hat{H}_L(z)$ at 0 follows from

$$\hat{H}_L(0) = \lim_{x \to -\infty} \frac{(\hat{h}_L(x))^+}{\hat{G}(x)} = \lim_{x \to -\infty} \frac{0}{\hat{G}(x)} = 0,$$

$$\lim_{z \to 0} \hat{H}_L(z) = \lim_{x \to -\infty} \frac{\hat{h}_L}{\hat{G}}(x) = \lim_{x \to -\infty} \frac{-(c_s + c_b)}{\hat{G}(x)} = 0,$$

where we have used that $z = \hat{\psi}(x) \to 0$ as $x \to -\infty$.

Furthermore, for $x \in (-\infty, L] \cup [b_L^*, +\infty)$, we have $V_L(x) = x - c_s$, and thus, $\hat{h}_L(x) = -(c_s + c_b)$. Also, with the facts that $\hat{\psi}(x)$ is a strictly increasing function and $\hat{G}(x) > 0$, property (i) follows.

(ii) By the definition of $\hat{H}_L$, since $\hat{G}$ and $\hat{\psi}$ are differentiable everywhere, we only need to show the differentiability of $V_L(x)$. To this end, $V_L(x)$ is differentiable at b_L^* by (2.44)-(2.45), but not at L. Therefore, $\hat{H}_L$ is differentiable for $z \in (0, \hat{\psi}(L)) \cup (\hat{\psi}(L), +\infty)$.

In view of the facts that $\hat{G}'(x) < 0$, $\hat{\psi}'(x) > 0$, and $\hat{G}^2(x) > 0$, we have for $x \in (-\infty, L) \cup [b_L^*, +\infty)$,

$$\hat{H}_L'(z) = \frac{1}{\hat{\psi}'(x)} \left(\frac{\hat{h}_L}{\hat{G}}\right)'(x) = \frac{1}{\hat{\psi}'(x)} \left(\frac{-(c_s + c_b)}{\hat{G}(x)}\right)' = \frac{(c_s + c_b)\hat{G}'(x)}{\hat{\psi}'(x)\hat{G}^2(x)} < 0.$$

Therefore, $\hat{H}_L(z)$ is strictly decreasing for $z \in (0, \hat{\psi}(L)) \cup [\hat{\psi}(b_L^*), +\infty)$.

(iii) Both $\hat{G}$ and $\hat{\psi}$ are twice differentiable everywhere, while $V_L(x)$ is twice differentiable everywhere except at $x = L$ and b_L^*, and so is $\hat{h}_L(x)$. Therefore, $\hat{H}_L(z)$ is twice differentiable on $(0, \hat{\psi}(L)) \cup (\hat{\psi}(L), \hat{\psi}(b_L^*)) \cup (\hat{\psi}(b_L^*), +\infty)$.

To determine the convexity/concavity of $\hat{H}_L$, we look at the second order derivative:

$$\hat{H}_L''(z) = \frac{2}{\sigma^2 \hat{G}(x)(\hat{\psi}'(x))^2}(\mathcal{L} - \hat{r})\hat{h}_L(x),$$

whose sign is determined by

$$(\mathcal{L} - \hat{r})\hat{h}_L(x)$$
$$= \frac{1}{2}\sigma^2 V_L''(x) + \mu(\theta - x)V_L'(x) - \mu(\theta - x) - \hat{r}(V_L(x) - x - c_b)$$
$$= \begin{cases} (r - \hat{r})V_L(x) + (\mu + \hat{r})x - \mu\theta + \hat{r}c_b & \text{if } x \in (L, b_L^*), \\ \hat{r}(c_s + c_b) > 0 & \text{if } x \in (-\infty, L) \cup (b_L^*, +\infty). \end{cases}$$

This implies that $\hat{H}_L$ is convex for $z \in (0, \hat{\psi}(L)) \cup (\hat{\psi}(b_L^*), +\infty)$.

On the other hand, the condition $\sup_{x\in\mathbb{R}} \hat{h}_L(x) > 0$ implies that

$$\sup_{z\in[0,+\infty)} \hat{H}_L(z) > 0.$$

By property (i) and twice differentiability of $\hat{H}_L(z)$ for $z \in (\hat{\psi}(L), \hat{\psi}(b_L^*))$, there must exist an interval $(\hat{\psi}(\underline{a}_L), \hat{\psi}(\bar{d}_L)) \subseteq (\hat{\psi}(L), \hat{\psi}(b_L^*))$ such that $\hat{H}_L(z)$ is concave, maximized at $\hat{z}_1 \in (\hat{\psi}(\underline{a}_L), \hat{\psi}(\bar{d}_L))$.

Furthermore, if $V_L(x)$ is strictly increasing on (L, b_L^*), then $(\mathcal{L} - \hat{r})\hat{h}_L(x)$ is also strictly increasing. To prove this, we first recall from Lemma 2.5 that $H(z)$ is strictly increasing and concave on $(\psi(L^*), +\infty)$. By Proposition 2.15, we have $b_L^* < b^*$, which implies $z_L < z^*$, and thus, $H'(z_L) > H'(z^*)$.

Then, it follows from (2.32), (2.33) and (2.47) that

$$W_L'(z) = H'(z_L) > H'(z^*) = W'(z), \quad \text{for } z \in (\psi(L), z_L).$$

Next, since

$$W_L(z) = \frac{V_L}{G} \circ \psi^{-1}(z),$$

we differentiate to get

$$W_L'(z) = \frac{1}{\psi'(x)} (\frac{V_L}{G})'(x) = \frac{1}{\psi'(x)} (\frac{V_L'(x)G(x) - V_L(x)G'(x)}{G^2(x)}).$$

The same holds for $W'(z)$ with $V(x)$ replacing $V_L(x)$. As both $\psi'(x)$ and $G^2(x)$ are positive, $W_L'(z) > W'(z)$ is equivalent to $V_L'(x)G(x) - V_L(x)G'(x) > V'(x)G(x) - V(x)G'(x)$. This implies that

$$V_L'(x) - V'(x) = -\frac{G'(x)}{G(x)}(V(x) - V_L(x)) > 0,$$

since $G(x) > 0$, $G'(x) < 0$, and $V(x) > V_L(x)$. Recalling that $V'(x) > 0$, we have established that $V_L(x)$ is a strictly increasing function, and so is $(\mathcal{L} - \hat{r})\hat{h}_L(x)$. As we have shown the existence of an interval $(\hat{\psi}(\underline{a}_L), \hat{\psi}(\bar{d}_L)) \subseteq (\hat{\psi}(L), \hat{\psi}(b_L^*))$ over which $\hat{H}(z)$ is concave, or equivalently $(\mathcal{L} - \hat{r})\hat{h}_L(x) < 0$ with $x = \hat{\psi}^{-1}(z)$. Then by the strictly increasing property of $(\mathcal{L} - \hat{r})\hat{h}_L(x)$, we conclude $\underline{a}_L = L$ and $\bar{d}_L \in (L, b_L^*)$ is the unique solution to $(\mathcal{L} - \hat{r})\hat{h}_L(x) = 0$, and

$$(\mathcal{L} - \hat{r})\hat{h}_L(x) \begin{cases} < 0 & \text{if } x \in (L, \bar{d}_L), \\ > 0 & \text{if } x \in (-\infty, L) \cup (\bar{d}_L, b_L^*) \cup (b_L^*, +\infty). \end{cases}$$

Hence, we conclude the convexity and concavity of the function $\hat{H}_L$.

Chapter 3

Trading Under the Exponential OU Model

Another widely used mean-reverting process is the exponential Ornstein-Uhlenbeck (XOU) process:

$$\xi_t = e^{X_t}, \qquad t \geq 0, \tag{3.1}$$

where X is the OU process defined in (2.1). In other words, X is the log-price of the positive XOU process ξ. In this chapter, we solve an optimal double stopping problem to determine the optimal times to enter and subsequently exit the market, when prices are driven by an exponential Ornstein-Uhlenbeck process. In addition, we analyze a related optimal switching problem that involves an infinite sequence of trades. Among our results, we find that the investor generally enters when the price is low, but may find it optimal to wait if the current price is sufficiently close to zero. In other words, the continuation (waiting) region for entry is *disconnected*. Numerical results are provided to illustrate the dependence of timing strategies on model parameters and transaction costs.

In Section 3.1, we formulate both the optimal double stopping and optimal switching problems. Then, we present our analytical and numerical results in Section 3.2. The proofs of our main results are detailed in Sections 3.3 and 3.4.

3.1 Optimal Trading Problems

Given an XOU price process satisfying (3.1), we denote by $\mathbb{F}$ the filtration generated by the standard Brownian motion B (see (2.1)), and $\mathcal{T}$ the set of all $\mathbb{F}$-stopping times.

3.1.1 *Optimal Double Stopping Approach*

We first consider the optimal timing to sell. If the share of the asset is sold at some time τ, then the investor will receive the value $\xi_\tau = e^{X_\tau}$ and pay a constant transaction cost $c_s > 0$. To maximize the expected discounted value, the investor solves the optimal stopping problem

$$V^\xi(x) = \sup_{\tau \in \mathcal{T}} \mathbb{E}_x \left\{ e^{-r\tau}(e^{X_\tau} - c_s) \right\}, \tag{3.2}$$

where $r > 0$ is the constant discount rate, and $\mathbb{E}_x\{\cdot\} \equiv \mathbb{E}\{\cdot | X_0 = x\}$.

The value function $V^\xi(x)$ represents the expected liquidation value associated with ξ. On the other hand, the current price plus the transaction cost constitute the total cost to enter the trade. Before even holding the risky asset, the investor can always choose the optimal timing to start the trade, or not to enter at all. This leads us to analyze the entry timing inherent in the trading problem. Precisely, we solve

$$J^\xi(x) = \sup_{\nu \in \mathcal{T}} \mathbb{E}_x \left\{ e^{-r\nu}(V^\xi(X_\nu) - e^{X_\nu} - c_b) \right\}, \tag{3.3}$$

with the constant transaction cost $c_b > 0$ incurred at the time of purchase. In other words, the trader seeks to maximize the expected difference between the value function $V^\xi(X_\nu)$ and the current e^{X_ν}, minus transaction cost c_b. The value function $J^\xi(x)$ represents the maximum expected value of the investment opportunity in the price process ξ, with transaction costs c_b and c_s incurred, respectively, at entry and exit. For our analysis, the transaction costs c_b and c_s can be different. To facilitate presentation, we denote the functions

$$h_s^\xi(x) = e^x - c_s \quad \text{and} \quad h_b^\xi(x) = e^x + c_b. \tag{3.4}$$

If it turns out that $J^\xi(X_0) \le 0$ for some initial value X_0, then the investor will not start to trade X. In view of Example 2.1, it is important to identify the trivial cases under any given dynamics. Under the XOU model, since $\sup_{x \in \mathbb{R}}(V^\xi(x) - h_b^\xi(x)) \le 0$ implies that $J^\xi(x) \le 0$ for $x \in \mathbb{R}$, we shall therefore focus on the case with

$$\sup_{x \in \mathbb{R}}(V^\xi(x) - h_b^\xi(x)) > 0, \tag{3.5}$$

and solve for the non-trivial optimal timing strategy.

3.1.2 *Optimal Switching Approach*

Under the optimal switching approach, the investor is assumed to commit to an infinite number of trades. The sequential trading times are modeled by the stopping times $\nu_1, \tau_1, \nu_2, \tau_2, \cdots \in \mathcal{T}$ such that

$$0 \le \nu_1 \le \tau_1 \le \nu_2 \le \tau_2 \le \ldots.$$

A share of the risky asset is bought and sold, respectively, at times ν_i and τ_i, $i \in \mathbb{N}$. The investor's optimal timing to trade would depend on the initial position. Precisely, under the XOU model, if the investor starts with a zero position, then the first trading decision is when to *buy* and the corresponding optimal switching problem is

$$\tilde{J}^{\xi}(x) = \sup_{\Lambda_0} \mathbb{E}_x \left\{ \sum_{n=1}^{\infty} [e^{-r\tau_n} h_s^{\xi}(X_{\tau_n}) - e^{-r\nu_n} h_b^{\xi}(X_{\nu_n})] \right\}, \qquad (3.6)$$

with the set of admissible stopping times $\Lambda_0 = (\nu_1, \tau_1, \nu_2, \tau_2, \dots)$, and the reward functions h_s^{ξ} and h_b^{ξ} defined in (3.4). On the other hand, if the investor is initially holding a share of the asset, then the investor first determines when to *sell* and solves

$$\tilde{V}^{\xi}(x) = \sup_{\Lambda_1} \mathbb{E}_x \left\{ e^{-r\tau_1} h_s^{\xi}(X_{\tau_1}) + \sum_{n=2}^{\infty} [e^{-r\tau_n} h_s^{\xi}(X_{\tau_n}) - e^{-r\nu_n} h_b^{\xi}(X_{\nu_n})] \right\},$$
$$(3.7)$$

with $\Lambda_1 = (\tau_1, \nu_2, \tau_2, \nu_3, \dots)$.

In summary, the optimal double stopping and switching problems differ in the number of trades. Observe that any strategy for the double stopping problems (3.2) and (3.3) are also candidate strategies for the switching problems (3.7) and (3.6) respectively. Therefore, it follows that $V^{\xi}(x) \le \tilde{V}^{\xi}(x)$ and $J^{\xi}(x) \le \tilde{J}^{\xi}(x)$. Our objective is to derive and compare the corresponding optimal timing strategies under these two approaches.

3.2 Summary of Analytical Results

We first summarize our analytical results and illustrate the optimal trading strategies. The method of solutions and their proofs will be discussed in Section 3.3. We begin with the optimal stopping problems (3.2) and (3.3) under the XOU model.

3.2.1 *Optimal Double Stopping Problem*

We now present the result for the optimal exit timing problem under the XOU model. First, we obtain a bound for the value function V^ξ.

Lemma 3.1. *There exists a positive constant K^ξ such that, for all $x \in \mathbb{R}$,*
$0 \le V^\xi(x) \le e^x + K^\xi$.

Theorem 3.2. *The optimal liquidation problem* (3.2) *admits the solution*

$$V^\xi(x) = \begin{cases} \frac{e^{b^{\xi*}} - c_s}{F(b^{\xi*})} F(x) & \text{if } x < b^{\xi*}, \\ e^x - c_s & \text{if } x \ge b^{\xi*}, \end{cases} \tag{3.8}$$

where the optimal log-price level $b^{\xi*}$ for liquidation is uniquely found from the equation

$$e^b F(b) = (e^b - c_s) F'(b). \tag{3.9}$$

The optimal liquidation time is given by

$$\tau^{\xi*} = \inf\{\, t \ge 0 \,:\, X_t \ge b^{\xi*} \,\} = \inf\{\, t \ge 0 \,:\, \xi_t \ge e^{b^{\xi*}} \,\}.$$

We now turn to the optimal entry timing problem, and give a bound on the value function J^ξ.

Lemma 3.3. *There exists a positive constant $\hat{K}^\xi$ such that, for all $x \in \mathbb{R}$,*
$0 \le J^\xi(x) \le \hat{K}^\xi$.

Theorem 3.4. *Under the XOU model, the optimal entry timing problem* (3.3) *admits the solution*

$$J^\xi(x) = \begin{cases} P^\xi F(x) & \text{if } x \in (-\infty, a^{\xi*}), \\ V^\xi(x) - (e^x + c_b) & \text{if } x \in [a^{\xi*}, d^{\xi*}], \\ Q^\xi G(x) & \text{if } x \in (d^{\xi*}, +\infty), \end{cases} \tag{3.10}$$

with the constants

$$P^\xi = \frac{V^\xi(a^{\xi*}) - (e^{a^{\xi*}} + c_b)}{F(a^{\xi*})}, \quad Q^\xi = \frac{V^\xi(d^{\xi*}) - (e^{d^{\xi*}} + c_b)}{G(d^{\xi*})},$$

and the critical levels $a^{\xi*}$ and $d^{\xi*}$ satisfying, respectively,

$$F(a)(V^{\xi'}(a) - e^a) = F'(a)(V^\xi(a) - (e^a + c_b)), \tag{3.11}$$

$$G(d)(V^{\xi'}(d) - e^d) = G'(d)(V^\xi(d) - (e^d + c_b)). \tag{3.12}$$

The optimal entry time is given by

$$\nu_{a^{\xi*},d^{\xi*}} := \inf\{\, t \ge 0 \,:\, X_t \in [a^{\xi*}, d^{\xi*}] \,\}.$$

In summary, the investor should exit the market as soon as the price reaches the upper level $e^{b^{\xi^*}}$. In contrast, the optimal entry timing is the first time that the XOU price ξ enters the interval $[e^{a^{\xi^*}}, e^{d^{\xi^*}}]$. In other words, it is optimal to wait if the current price ξ_t is too close to zero, i.e. if $\xi_t < e^{a^{\xi^*}}$. Moreover, the interval $[e^{a^{\xi^*}}, e^{d^{\xi^*}}]$ is contained in $(0, e^{b^{\xi^*}})$, and thus, the continuation region for market entry is *disconnected*. One reason for this phenomenon is that waiting to enter the market helps reduce the effective transaction cost due to discounting. This effect outweighs the spread in values between the value function and current asset value in this case.

3.2.2 *Optimal Switching Problem*

We now turn to the optimal switching problems defined in (3.6) and (3.7) under the XOU model. To facilitate the presentation, we denote

$$f_s(x) := (\mu\theta + \frac{1}{2}\sigma^2 - r) - \mu x + r c_s e^{-x},$$

$$f_b(x) := (\mu\theta + \frac{1}{2}\sigma^2 - r) - \mu x - r c_b e^{-x}.$$

Applying the operator $\mathcal{L}$ (see (2.9)) to h_s^ξ and h_b^ξ (see (3.4)), it follows that $(\mathcal{L} - r)h_s^\xi(x) = e^x f_s(x)$ and $(\mathcal{L} - r)h_b^\xi(x) = e^x f_b(x)$. Therefore, f_s (resp. f_b) preserves the sign of $(\mathcal{L} - r)h_s^\xi$ (resp. $(\mathcal{L} - r)h_b^\xi$). It can be shown that $f_s(x) = 0$ has a unique root, denoted by x_s. However,

$$f_b(x) = 0 \tag{3.13}$$

may have no root, a single root, or two distinct roots, denoted by x_{b1} and x_{b2}, if they exist. The following observations will also be useful:

$$f_s(x) \begin{cases} > 0 & \text{if } x < x_s, \\ < 0 & \text{if } x > x_s, \end{cases} \quad \text{and} \quad f_b(x) \begin{cases} < 0 & \text{if } x \in (-\infty, x_{b1}) \cup (x_{b2}, +\infty), \\ > 0 & \text{if } x \in (x_{b1}, x_{b2}). \end{cases}$$

$$\tag{3.14}$$

We first obtain bounds for the value functions $\tilde{J}^\xi$ and $\tilde{V}^\xi$.

Lemma 3.5. *There exists positive constants C_1 and C_2 such that*

$$0 \le \tilde{J}^\xi(x) \le C_1,$$

$$0 \le \tilde{V}^\xi(x) \le e^x + C_2.$$

The optimal switching problems have two different sets of solutions depending on the problem data.

Theorem 3.6. *The optimal switching problem defined in (3.6)–(3.7) admits the solution*

$$\tilde{J}^{\xi}(x) = 0, \ for \ x \in \mathbb{R}, \quad and \quad \tilde{V}^{\xi}(x) = \begin{cases} \frac{e^{b^{\xi*}} - c_s}{F(b^{\xi*})} F(x) & if \ x < b^{\xi*}, \\ e^x - c_s & if \ x \geq b^{\xi*}, \end{cases}$$

$$(3.15)$$

where $b^{\xi*}$ satisfies (3.9), if any of the following mutually exclusive conditions holds:

(i) There is no root or a single root to equation (3.13).
(ii) There are two distinct roots to (3.13). Also

$$\exists \, \tilde{a}^* \in (x_{b1}, x_{b2}) \quad such \ that \quad F(\tilde{a}^*)e^{\tilde{a}^*} = F'(\tilde{a}^*)(e^{\tilde{a}^*} + c_b), \ (3.16)$$

and

$$\frac{e^{\tilde{a}^*} + c_b}{F(\tilde{a}^*)} \geq \frac{e^{b^{\xi*}} - c_s}{F(b^{\xi*})}. \tag{3.17}$$

(iii) There are two distinct roots to (3.13) but (3.16) does not hold.

In Theorem 3.6, $\tilde{J}^{\xi} = 0$ means that it is optimal not to enter the market at all. On the other hand, if one starts with a unit of the underlying asset, the optimal switching problem reduces to a problem of optimal single stopping. Indeed, the investor will never re-enter the market after exit. This is identical to the optimal liquidation problem (3.2) where there is only a single (exit) trade. The optimal strategy in this case is the same as V^{ξ} in (3.8) – it is optimal to exit the market as soon as the log-price X reaches the threshold $b^{\xi*}$.

We also address the remaining case when none of the conditions in Theorem 3.6 hold. As we show next, the optimal strategy will involve both entry and exit thresholds.

Theorem 3.7. *If there are two distinct roots to (3.13), x_{b1} and x_{b2}, and there exists a number $\tilde{a}^* \in (x_{b1}, x_{b2})$ satisfying (3.16) such that*

$$\frac{e^{\tilde{a}^*} + c_b}{F(\tilde{a}^*)} < \frac{e^{b^{\xi*}} - c_s}{F(b^{\xi*})}, \tag{3.18}$$

then the optimal switching problem (3.6)-(3.7) *admits the solution*

$$\tilde{J}^{\xi}(x) = \begin{cases} \tilde{P}F(x) & \text{if } x \in (-\infty, \tilde{a}^*), \\ \tilde{K}F(x) - (e^x + c_b) & \text{if } x \in [\tilde{a}^*, \tilde{d}^*], \\ \tilde{Q}G(x) & \text{if } x \in (\tilde{d}^*, +\infty), \end{cases} \qquad (3.19)$$

$$\tilde{V}^{\xi}(x) = \begin{cases} \tilde{K}F(x) & \text{if } x \in (-\infty, \tilde{b}^*), \\ \tilde{Q}G(x) + e^x - c_s & \text{if } x \in [\tilde{b}^*, +\infty), \end{cases} \qquad (3.20)$$

where $\tilde{a}^*$ *satisfies* (3.16), *and*

$$\tilde{P} = \tilde{K} - \frac{e^{\tilde{a}^*} + c_b}{F(\tilde{a}^*)},$$

$$\tilde{K} = \frac{e^{\tilde{d}^*} G(\tilde{d}^*) - (e^{\tilde{d}^*} + c_b) G'(\tilde{d}^*)}{F'(\tilde{d}^*) G(\tilde{d}^*) - F(\tilde{d}^*) G'(\tilde{d}^*)},$$

$$\tilde{Q} = \frac{e^{\tilde{d}^*} F(\tilde{d}^*) - (e^{\tilde{d}^*} + c_b) F'(\tilde{d}^*)}{F'(\tilde{d}^*) G(\tilde{d}^*) - F(\tilde{d}^*) G'(\tilde{d}^*)},$$

There exist unique critical levels $\tilde{d}^*$ *and* $\tilde{b}^*$ *which are found from the non-linear system of equations:*

$$\frac{e^d G(d) - (e^d + c_b) G'(d)}{F'(d) G(d) - F(d) G'(d)} = \frac{e^b G(b) - (e^b - c_s) G'(b)}{F'(b) G(b) - F(b) G'(b)}, \qquad (3.21)$$

$$\frac{e^d F(d) - (e^d + c_b) F'(d)}{F'(d) G(d) - F(d) G'(d)} = \frac{e^b F(b) - (e^b - c_s) F'(b)}{F'(b) G(b) - F(b) G'(b)}. \qquad (3.22)$$

Moreover, the critical levels are such that $\tilde{d}^* \in (x_{b1}, x_{b2})$ *and* $\tilde{b}^* > x_s$.

The optimal strategy in Theorem 3.7 is described by the stopping times $\Lambda_0^* = (\nu_1^*, \tau_1^*, \nu_2^*, \tau_2^*, \dots)$, and $\Lambda_1^* = (\tau_1^*, \nu_2^*, \tau_2^*, \nu_3^*, \dots)$, with

$$\nu_1^* = \inf\{t \geq 0 : X_t \in [\tilde{a}^*, \tilde{d}^*]\},$$

$$\tau_i^* = \inf\{t \geq \nu_i^* : X_t \geq \tilde{b}^*\}, \quad \text{and} \quad \nu_{i+1}^* = \inf\{t \geq \tau_i^* : X_t \leq \tilde{d}^*\}, \quad \text{for } i \geq 1.$$

In other words, it is optimal to buy if the price is within $[e^{\tilde{a}^*}, e^{\tilde{d}^*}]$ and then sell when the price ξ reaches $e^{\tilde{b}^*}$. The structure of the buy/sell regions is similar to that in the double stopping case (see Theorems 3.2 and 3.4). In particular, $\tilde{a}^*$ is the same as $a^{\xi*}$ in Theorem 3.4 since the equations (3.11) and (3.16) are equivalent. The level $\tilde{a}^*$ is only relevant to the first purchase. Mathematically, $\tilde{a}^*$ is determined separately from $\tilde{d}^*$ and $\tilde{b}^*$. If we start with a zero position, then it is optimal to enter if the price ξ lies in the interval $[e^{\tilde{a}^*}, e^{\tilde{d}^*}]$. However, on all subsequent trades, we enter as soon as the price hits $e^{\tilde{d}^*}$ from above (after exiting at $e^{\tilde{b}^*}$ previously). Hence, the lower level $\tilde{a}^*$ becomes irrelevant after the first entry.

Note that the conditions that differentiate Theorems 3.6 and 3.7 are exhaustive and mutually exclusive. If the conditions in Theorem 3.6 are violated, then the conditions in Theorem 3.7 must hold. In particular, condition (3.16) in Theorem 3.6 holds if and only if

$$\left| \int_{-\infty}^{x_{b1}} \Psi(x) e^x f_b(x) dx \right| < \int_{x_{b1}}^{x_{b2}} \Psi(x) e^x f_b(x) dx, \qquad (3.23)$$

where

$$\Psi(x) = \frac{2F(x)}{\sigma^2 \mathcal{W}(x)}, \quad \text{and} \quad \mathcal{W}(x) = F'(x)G(x) - F(x)G'(x) > 0. \quad (3.24)$$

Inequality (3.23) can be numerically verified given the model inputs.

3.2.3   *Numerical Examples*

We numerically implement Theorems 3.2, 3.4, and 3.7, and illustrate the associated entry/exit thresholds. In Figure 3.1(a), the optimal entry levels $d^{\xi *}$ and $\tilde{d}^*$ rise, respectively, from 0.7425 to 0.7912 and from 0.8310 to 0.8850, as the speed of mean reversion μ increases from 0.5 to 1. On the other hand, the critical exit levels $b^{\xi *}$ and $\tilde{b}^*$ remain relatively flat over μ. As for the critical lower level $a^{\xi *}$ from the optimal double stopping problem, Figure 3.1(b) shows that it is decreasing in μ. The same pattern holds for the optimal switching problem since the critical lower level $\tilde{a}^*$ is identical to $a^{\xi *}$, as noted above.

We now look at the impact of transaction cost in Figure 3.2. On the left panel, we observe that as the transaction cost c_b increases, the gap between the optimal switching entry and exit levels, $\tilde{d}^*$ and $\tilde{b}^*$, widens. This means that it is optimal to delay both entry and exit. Intuitively, to counter the fall in profit margin due to an increase in transaction cost, it is necessary to buy at a lower price and sell at a higher price to seek a wider spread. In comparison, the exit level $b^{\xi *}$ from the double stopping problem is known analytically to be independent of the entry cost, so it stays constant as c_b increases in the figure. In contrast, the entry level $d^{\xi *}$, however, decreases as c_b increases but much less significantly than $\tilde{d}^*$. Figure 3.2(b) shows that $a^{\xi *}$, which is the same for both the optimal double stopping and switching problems, increases monotonically with c_b.

In both Figures 3.1 and 3.2, we can see that the interval of the entry and exit levels, $(\tilde{d}^*, \tilde{b}^*)$, associated with the optimal switching problem lies

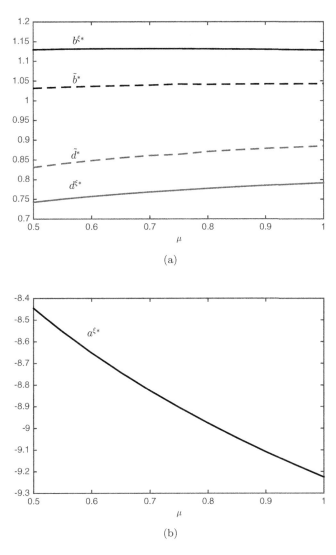

(a)

(b)

Fig. 3.1 (a) The optimal entry and exit levels vs speed of mean reversion μ. Parameters: $\sigma = 0.2$, $\theta = 1$, $r = 0.05$, $c_s = 0.02$, $c_b = 0.02$. (b) The critical lower level of entry region $a^{\xi*}$ decreases monotonically from -8.4452 to -9.2258 as μ increases from 0.5 to 1. Parameters: $\sigma = 0.2$, $\theta = 1$, $r = 0.05$, $c_s = 0.02$, $c_b = 0.02$.

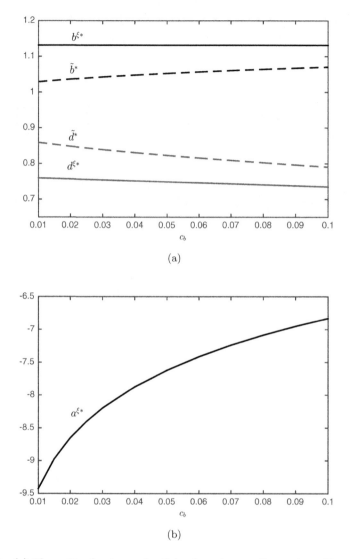

Fig. 3.2 (a) The optimal entry and exit levels vs transaction cost c_b. Parameters: $\mu = 0.6$, $\sigma = 0.2$, $\theta = 1$, $r = 0.05$, $c_s = 0.02$. (b) The critical lower level of entry region $a^{\xi*}$ increases monotonically from -9.4228 to -6.8305 as c_b increases from 0.01 to 0.1. Parameters: $\mu = 0.6$, $\sigma = 0.2$, $\theta = 1$, $r = 0.05$, $c_s = 0.02$.

within the corresponding interval $(d^{\xi *}, b^{\xi *})$ from the optimal double stopping problem. Intuitively, with the intention to enter the market again upon completing the current trade, the trader is more willing to enter/exit earlier, meaning a narrowed waiting region.

Figure 3.3 shows a simulated path and the associated entry/exit levels. As the path starts at $\xi_0 = 2.6011 > e^{\tilde{d}^*} > e^{d^{\xi *}}$, the investor waits to enter until the path reaches the lower level $e^{d^{\xi *}}$ (double stopping) or $e^{\tilde{d}^*}$ (switching) according to Theorems 3.4 and 3.7. After entry, the investor exits at the optimal level $e^{b^{\xi *}}$ (double stopping) or $e^{\tilde{b}^*}$ (switching). The optimal switching thresholds imply that the investor first enters the market on day 188 where the underlying asset price is 2.3847. In contrast, the optimal double stopping timing yields a later entry on day 845 when the price first reaches $e^{d^{\xi *}} = 2.1754$. As for the exit timing, under the optimal switching setting, the investor exits the market earlier on day 268 at the price $e^{\tilde{b}^*} = 2.8323$. The double stopping timing is much later on day 1160 when the price reaches $e^{b^{\xi *}} = 3.0988$. In addition, under the optimal switching problem, the investor executes more trades within the same time span. As seen in the figure, the investor would have completed two 'roundtrip' (buy-and-sell) trades in the market before the double stopping investor liquidates for the first time.

3.3 Methods of Solution

We now provide detailed proofs for our analytical results in Section 3.2 beginning with Theorems 3.2 and 3.4 for the optimal double stopping problems.

3.3.1 *Optimal Double Stopping Problem*

3.3.1.1 *Optimal Exit Timing*

To facilitate the presentation, we define the function H^ξ associated with the reward function h_s^ξ (see (3.4)) by following (2.17) for H (with h replaced by h^ξ). We summarize the functional properties of H^ξ.

Lemma 3.8. *The function H^ξ is continuous on $[0, +\infty)$, twice differentiable on $(0, +\infty)$ and possesses the following properties:*

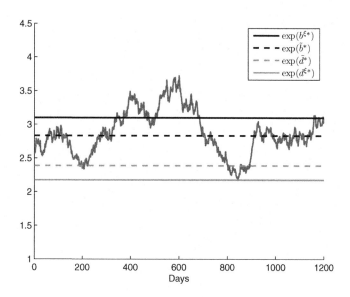

Fig. 3.3   A sample exponential OU path, along with entry and exit levels. Under the double stopping setting, the investor enters at $\nu_{d^{\xi*}} = \inf\{t \geq 0 : \xi_t \leq e^{d^{\xi*}} = 2.1754\}$ with $d^{\xi*} = 0.7772$, and exit at $\tau_{b^{\xi*}} = \inf\{t \geq \nu_{d^{\xi*}} : \xi_t \geq e^{b^{\xi*}} = 3.0988\}$ with $b^{\xi*} = 1.1310$. The optimal switching investor enters at $\nu_{\tilde{d}^*} = \inf\{t \geq 0 : \xi_t \leq e^{\tilde{d}^*} = 2.3888\}$ with $\tilde{d}^* = 0.8708$, and exit at $\tau_{\tilde{b}^*} = \inf\{t \geq \nu_{\tilde{d}^*} : \xi_t \geq e^{\tilde{b}^*} = 2.8323\}$ with $\tilde{b}^* = 1.0411$. The critical lower threshold of entry region is $e^{a^{\xi*}} = 1.264 \cdot 10^{-4}$ with $a^{\xi*} = -8.9760$ (not shown in this figure). Parameters: $\mu = 0.8$, $\sigma = 0.2$, $\theta = 1$, $r = 0.05$, $c_s = 0.02$, $c_b = 0.02$.

(i) $H^\xi(0) = 0$, and

$$H^{\xi'}(z) \begin{cases} < 0 & \text{if } z \in (0, \psi(\ln c_s)), \\ > 0 & \text{if } z \in (\psi(\ln c_s), +\infty). \end{cases}$$

(ii) $H^\xi(z)$ *is strictly increasing for* $z \in (\psi(\ln c_s), +\infty)$, *and* $H^{\xi'}(z) \to 0$ *as* $z \to +\infty$.

(iii)

$$H^\xi(z) \text{ is } \begin{cases} convex & \text{if } z \in (0, \psi(x_s)], \\ concave & \text{if } z \in [\psi(x_s), +\infty). \end{cases}$$

From Lemma 3.8, we see that H^ξ shares a very similar structure as H. Using the properties of H^ξ, we now solve for the optimal exit timing.

Proof of Theorem 3.2 We look for the value function of the form: $V^\xi(x) = G(x)W^\xi(\psi(x))$, where W^ξ is the smallest concave majorant of H^ξ. By Lemma 3.8, we deduce that H^ξ is concave over $[\psi(x_s), +\infty)$, strictly positive over $(\psi(\ln c_s), +\infty)$, and $H^{\xi\prime}(z) \to 0$ as $z \to +\infty$. Therefore, there exists a unique number $z^{\xi*} > \psi(x_s) \vee \psi(\ln c_s)$ such that

$$\frac{H^\xi(z^{\xi*})}{z^{\xi*}} = H^{\xi\prime}(z^{\xi*}). \tag{3.25}$$

In turn, the smallest concave majorant of H^ξ is given by

$$W^\xi(z) = \begin{cases} z \frac{H^\xi(z^{\xi*})}{z^{\xi*}} & \text{if } z \in [0, z^{\xi*}), \\ H^\xi(z) & \text{if } z \in [z^{\xi*}, +\infty). \end{cases}$$

Substituting $b^{\xi*} = \psi^{-1}(z^{\xi*})$ into (3.25), we have

$$\frac{H^\xi(z^{\xi*})}{z^{\xi*}} = \frac{H^\xi(\psi(b^{\xi*}))}{\psi(b^{\xi*})} = \frac{e^{b^{\xi*}} - c_s}{F(b^{\xi*})},$$

and

$$\begin{aligned} H^{\xi\prime}(z^{\xi*}) &= \frac{e^{\psi^{-1}(z^{\xi*})}G(\psi^{-1}(z^{\xi*})) - (e^{\psi^{-1}(z^{\xi*})} - c_s)G'(\psi^{-1}(z^{\xi*}))}{F'(\psi^{-1}(z^{\xi*}))G(\psi^{-1}(z^{\xi*})) - F(\psi^{-1}(z^{\xi*}))G'(\psi^{-1}(z^{\xi*}))} \\ &= \frac{e^{b^{\xi*}}G(b^{\xi*}) - (e^{b^{\xi*}} - c_s)G'(b^{\xi*})}{F'(b^{\xi*})G(b^{\xi*}) - F(b^{\xi*})G'(b^{\xi*})}. \end{aligned}$$

Equivalently, we can express (3.25) in terms of $b^{\xi*}$:

$$\frac{e^{b^{\xi*}} - c_s}{F(b^{\xi*})} = \frac{e^{b^{\xi*}}G(b^{\xi*}) - (e^{b^{\xi*}} - c_s)G'(b^{\xi*})}{F'(b^{\xi*})G(b^{\xi*}) - F(b^{\xi*})G'(b^{\xi*})},$$

which is equivalent to (3.9) after simplification. As a result, we have

$$W^\xi(\psi(x)) = \begin{cases} \psi(x)\frac{H^\xi(z^{\xi*})}{z^{\xi*}} = \frac{F(x)}{G(x)}\frac{e^{b^{\xi*}} - c_s}{F(b^{\xi*})} & \text{if } x \in (-\infty, b^{\xi*}), \\ H^\xi(\psi(x)) = \frac{e^x - c_s}{G(x)} & \text{if } x \in [b^{\xi*}, +\infty). \end{cases}$$

In turn, the value function $V^\xi(x) = G(x)W^\xi(\psi(x))$ is given by (3.8).

3.3.1.2 *Optimal Entry Timing*

We can directly follow the arguments that yield Theorem 2.3, but with the reward as $\hat{h}^\xi(x) = V^\xi(x) - h_b^\xi(x) = V^\xi(x) - (e^x + c_b)$ and define $\hat{H}^\xi$ analogous to H:

$$\hat{H}^\xi(z) := \begin{cases} \frac{\hat{h}^\xi}{G} \circ \psi^{-1}(z) & \text{if } z > 0, \\ \lim_{x \to -\infty} \frac{(\hat{h}^\xi(x))^+}{G(x)} & \text{if } z = 0. \end{cases}$$

We will look for the value function with the form: $J^\xi(x) = G(x)\hat{W}^\xi(\psi(x))$, where $\hat{W}^\xi$ is the smallest concave majorant of $\hat{H}^\xi$. The properties of $\hat{H}^\xi$ is given in the next lemma.

Lemma 3.9. *The function $\hat{H}^\xi$ is continuous on $[0, +\infty)$, differentiable on $(0, +\infty)$, and twice differentiable on $(0, \psi(b^{\xi*})) \cup (\psi(b^{\xi*}), +\infty)$, and possesses the following properties:*

(i) $\hat{H}^\xi(0) = 0$, *and there exists some* $\underline{b}^\xi < b^{\xi*}$ *such that* $\hat{H}^\xi(z) < 0$ *for* $z \in (0, \psi(\underline{b}^\xi)) \cup [\psi(b^{\xi*}), +\infty)$.

(ii) $\hat{H}^\xi(z)$ *is strictly decreasing for* $z \in [\psi(b^{\xi*}), +\infty)$.

(iii) *Define the constant*

$$x^{\xi*} = \theta + \frac{\sigma^2}{2\mu} - \frac{r}{\mu} - 1.$$

There exist some constants x_{b1} *and* x_{b2}, *with* $-\infty < x_{b1} < x^{\xi*} < x_{b2} < x_s$, *that solve* $f_b(x) = 0$, *such that*

$$\hat{H}^\xi(z) \text{ is } \begin{cases} convex & if\ y \in (0, \psi(x_{b1})) \cup (\psi(x_{b2}), +\infty) \\ concave & if\ z \in (\psi(x_{b1}), \psi(x_{b2})), \end{cases}$$

and $\hat{z}_1^\xi := \arg\max_{y \in [0, +\infty)} \hat{H}^\xi(y) \in (\psi(x_{b1}), \psi(x_{b2}))$.

Figure 3.4 gives a sketch of $\hat{H}^\xi$ according to Lemma 3.9, and illustrate the corresponding smallest concave majorant $\hat{W}^\xi$.

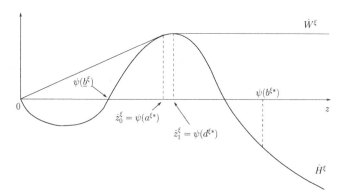

Fig. 3.4 Sketches of $\hat{H}^\xi$ and $\hat{W}^\xi$. The smallest concave majorant $\hat{W}^\xi$ is a straight line tangent to $\hat{H}^\xi$ at $\hat{z}_0^\xi$ on $[0, \hat{z}_0^\xi)$, coincides with $\hat{H}^\xi$ on $[\hat{z}_0^\xi, \hat{z}_1^\xi]$, and is equal to $\hat{H}^\xi(\hat{z}_1^\xi)$ on $(\hat{z}_1^\xi, +\infty)$.

Proof of Theorem 3.4 As in Lemma 3.9 and Figure 3.4, by the definition of the maximizer of $\hat{H}^\xi$, $\hat{z}_1^\xi$ satisfies the equation

$$\hat{H}^{\xi'}(\hat{z}_1^\xi) = 0. \tag{3.26}$$

Also there exists a unique number $\hat{z}_0^\xi \in (x_{b1}, \hat{z}_1^\xi)$ such that

$$\frac{\hat{H}^\xi(\hat{z}_0^\xi)}{\hat{z}_0^\xi} = \hat{H}^{\xi'}(\hat{z}_0^\xi). \tag{3.27}$$

Using (3.26), (3.27) and Figure 3.4, $\hat{W}^\xi$ is a straight line tangent to $\hat{H}^\xi$ at $\hat{z}_0^\xi$ on $[0, \hat{z}_0^\xi)$, coincides with $\hat{H}^\xi$ on $[\hat{z}_0^\xi, \hat{z}_1^\xi]$, and is equal to $\hat{H}^\xi(\hat{z}_1^\xi)$ on $(\hat{z}_1^\xi, +\infty)$. As a result,

$$\hat{W}^\xi(z) = \begin{cases} z\hat{H}^{\xi'}(\hat{z}_0^\xi) & \text{if } z \in [0, \hat{z}_0^\xi), \\ \hat{H}^\xi(z) & \text{if } z \in [\hat{z}_0^\xi, \hat{z}_1^\xi], \\ \hat{H}^\xi(\hat{z}_1^\xi) & \text{if } z \in (\hat{z}_1^\xi, +\infty). \end{cases}$$

Substituting $a^{\xi*} = \psi^{-1}(\hat{z}_0^\xi)$ into (3.27), we have

$$\frac{\hat{H}^\xi(\hat{z}_0^\xi)}{\hat{z}_0^\xi} = \frac{V^\xi(a^{\xi*}) - (e^{a^{\xi*}} + c_b)}{F(a^{\xi*})},$$

and

$$\hat{H}^{\xi'}(\hat{z}_0^\xi) = \frac{G(a^{\xi*})(V^{\xi'}(a^{\xi*}) - e^{a^{\xi*}}) - G'(a^{\xi*})(V^\xi(a^{\xi*}) - (e^{a^{\xi*}} + c_b))}{F'(a^{\xi*})G(a^{\xi*}) - F(a^{\xi*})G'(a^{\xi*})}.$$

Equivalently, we can express condition (3.27) in terms of $a^{\xi*}$:

$$\frac{V^\xi(a^{\xi*}) - (e^{a^{\xi*}} + c_b)}{F(a^{\xi*})}$$

$$= \frac{G(a^{\xi*})(V^{\xi'}(a^{\xi*}) - e^{a^{\xi*}}) - G'(a^{\xi*})(V^\xi(a^{\xi*}) - (e^{a^{\xi*}} + c_b))}{F'(a^{\xi*})G(a^{\xi*}) - F(a^{\xi*})G'(a^{\xi*})},$$

which is equivalent to (3.11) after simplification. Also, we can express $\hat{H}^{\xi'}(\hat{z}_0^\xi)$ in terms of $a^{\xi*}$:

$$\hat{H}^{\xi'}(\hat{z}_0^\xi) = \frac{\hat{H}^\xi(\hat{z}_0^\xi)}{\hat{z}_0^\xi} = \frac{V^\xi(a^{\xi*}) - (e^{a^{\xi*}} + c_b)}{F(a^{\xi*})} = P^\xi.$$

In addition, substituting $d^{\xi*} = \psi^{-1}(\hat{z}_1^\xi)$ into (3.26), we have

$$\frac{G(d^{\xi*})(V^{\xi'}(d^{\xi*}) - e^{d^{\xi*}}) - G'(d^{\xi*})(V^\xi(d^{\xi*}) - (e^{d^{\xi*}} + c_b))}{F'(d^{\xi*})G(d^{\xi*}) - F(d^{\xi*})G'(d^{\xi*})} = 0,$$

which can be further simplified to (3.12). Moreover, $\hat{H}^\xi(\hat{z}_1^\xi)$ can be written in terms of $d^{\xi*}$:

$$\hat{H}^\xi(\hat{z}_1^\xi) = \frac{V^\xi(d^{\xi*}) - (e^{d^{\xi*}} + c_b)}{G(d^{\xi*})} = Q^\xi.$$

By direct substitution of the expressions for $\hat{W}^\xi$ and the associated functions, we obtain the value function in (3.10).

3.3.2   *Optimal Switching Problem*

Using the results derived in previous sections, we can infer the structure of
the buy and sell regions of the switching problem and then proceed to verify
its optimality. In this section, we provide detailed proofs for Theorems 3.6
and 3.7.

Proof of Theorem 3.6 (Part 1)   First, with $h_s^\xi(x) = e^x - c_s$, we differentiate to get

$$\left(\frac{h_s^\xi}{F}\right)'(x) = \frac{(e^x - c_s)F'(x) - e^x F(x)}{F^2(x)}. \tag{3.28}$$

On the other hand, by Ito's lemma, we have

$$h_s^\xi(x) = \mathbb{E}_x\{e^{-rt}h_s^\xi(X_t)\} - \mathbb{E}_x\left\{\int_0^t e^{-ru}(\mathcal{L} - r)h_s^\xi(X_u)du\right\}.$$

Note that

$$\mathbb{E}_x\{e^{-rt}h_s^\xi(X_t)\} = e^{-rt}\left(e^{(x-\theta)e^{-\mu t}+\theta+\frac{\sigma^2}{4\mu}(1-e^{-2\mu t})} - c_s\right) \to 0 \quad \text{as} \quad t \to +\infty.$$

This implies that

$$h_s^\xi(x) = -\mathbb{E}_x\left\{\int_0^{+\infty} e^{-ru}(\mathcal{L} - r)h_s^\xi(X_u)du\right\}$$

$$= -G(x)\int_{-\infty}^x \Psi(s)(\mathcal{L} - r)h_s^\xi(s)ds$$

$$\quad - F(x)\int_x^{+\infty} \Phi(s)(\mathcal{L} - r)h_s^\xi(s)ds, \tag{3.29}$$

where Ψ is defined in (3.24) and

$$\Phi(x) := \frac{2G(x)}{\sigma^2 \mathcal{W}(x)}. \tag{3.30}$$

The last line follows from Theorem 50.7 in Rogers and Williams (2000,
p. 293). Dividing both sides by $F(x)$ and differentiating the RHS of (3.29),
we obtain

$$\left(\frac{h_s^\xi}{F}\right)'(x) = -\left(\frac{G}{F}\right)'(x)\int_{-\infty}^x \Psi(s)(\mathcal{L} - r)h_s^\xi(s)ds$$

$$\quad - \frac{G}{F}(x)\Psi(x)(\mathcal{L} - r)h_s^\xi(x) - \Phi(x)(\mathcal{L} - r)h_s^\xi(x)$$

$$= \frac{\mathcal{W}(x)}{F^2(x)}\int_{-\infty}^x \Psi(s)(\mathcal{L} - r)h_s^\xi(s)ds = \frac{\mathcal{W}(x)}{F^2(x)}q(x),$$

where

$$q(x) := \int_{-\infty}^{x} \Psi(s)(\mathcal{L} - r)h_s^{\xi}(s)ds.$$

Since $\mathcal{W}(x), F(x) > 0$, we deduce that $\left(\frac{h_s^{\xi}}{F}\right)'(x) = 0$ is equivalent to $q(x) = 0$. Using (3.28), we now see that (3.9) is equivalent to $q(b) = 0$.

Next, it follows from (3.14) that

$$q'(x) = \Psi(x)(\mathcal{L} - r)h_s^{\xi}(x) \begin{cases} > 0 & \text{if } x < x_s, \\ < 0 & \text{if } x > x_s. \end{cases} \tag{3.31}$$

This, together with the fact that $\lim_{x \to -\infty} q(x) = 0$, implies that there exists a unique $b^{\xi*}$ such that $q(b^{\xi*}) = 0$ if and only if $\lim_{x \to +\infty} q(x) < 0$. Next, we show that this inequality holds. By the definition of h_s^{ξ} and F, we have

$$\frac{h_s^{\xi}(x)}{F(x)} = \frac{e^x - c_s}{F(x)} > 0 \quad \text{for } x > \ln c_s, \qquad \lim_{x \to +\infty} \frac{h_s^{\xi}(x)}{F(x)} = 0,$$

$$\left(\frac{h_s^{\xi}}{F}\right)'(x) = \frac{\mathcal{W}(x)}{F^2(x)} \int_{-\infty}^{x} \Psi(s)(\mathcal{L} - r)h_s^{\xi}(s)ds = \frac{\mathcal{W}(x)}{F^2(x)}q(x). \tag{3.32}$$

Since q is strictly decreasing in $(x_s, +\infty)$, the above hold true if and only if $\lim_{x \to +\infty} q(x) < 0$. Therefore, we conclude that there exits a unique $b^{\xi*}$ such that $e^b F(b) = (e^b - c_s)F'(b)$. Using (3.31), we see that

$$b^{\xi*} > x_s \quad \text{and} \quad q(x) > 0 \quad \text{for all} \quad x < b^{\xi*}. \tag{3.33}$$

Observing that $e^{b^{\xi*}}, F(b^{\xi*}), F'(b^{\xi*}) > 0$, we can conclude that $h_s^{\xi}(b^{\xi*}) = e^{b^{\xi*}} - c_s > 0$, or equivalently $b^{\xi*} > \ln c_s$.

We now verify by direct substitution that $\tilde{V}^{\xi}(x)$ and $\tilde{J}^{\xi}(x)$ in (3.15) satisfy the pair of variational inequalities:

$$\min\{r\tilde{J}^{\xi}(x) - \mathcal{L}\tilde{J}^{\xi}(x), \tilde{J}^{\xi}(x) - (\tilde{V}^{\xi}(x) - h_b^{\xi}(x))\} = 0, \tag{3.34}$$

$$\min\{r\tilde{V}^{\xi}(x) - \mathcal{L}\tilde{V}^{\xi}(x), \tilde{V}^{\xi}(x) - (\tilde{J}^{\xi}(x) + h_s^{\xi}(x))\} = 0. \tag{3.35}$$

First, note that $\tilde{J}^{\xi}(x)$ is identically 0 and thus satisfies the equality

$$(r - \mathcal{L})\tilde{J}^{\xi}(x) = 0. \tag{3.36}$$

To show that $\tilde{J}^{\xi}(x) - (\tilde{V}^{\xi}(x) - h_b^{\xi}(x)) \geq 0$, we look at the disjoint intervals $(-\infty, b^{\xi*})$ and $[b^{\xi*}, \infty)$ separately. For $x \geq b^{\xi*}$, we have

$$\tilde{V}^{\xi}(x) - h_b^{\xi}(x) = -(c_b + c_s),$$

which implies $\tilde{J}^\xi(x) - (\tilde{V}^\xi(x) - h_b^\xi(x)) = c_b + c_s \geq 0$. When $x < b^{\xi*}$, the inequality

$$\tilde{J}^\xi(x) - (\tilde{V}^\xi(x) - h_b^\xi(x)) \geq 0$$

can be rewritten as

$$\frac{h_b^\xi(x)}{F(x)} = \frac{e^x + c_b}{F(x)} \geq \frac{e^{b^{\xi*}} - c_s}{F(b^{\xi*})} = \frac{h_s^\xi(b^{\xi*})}{F(b^{\xi*})}. \tag{3.37}$$

To determine the necessary conditions for this to hold, we consider the derivative of the LHS of (3.37):

$$\left(\frac{h_b^\xi}{F}\right)'(x) = \frac{W(x)}{F^2(x)} \int_{-\infty}^x \Psi(s)(\mathcal{L} - r)h_b^\xi(s)ds \tag{3.38}$$

$$= \frac{W(x)}{F^2(x)} \int_{-\infty}^x \Psi(s)e^s f_b(s)ds.$$

If $f_b(x) = 0$ has no roots, then $(\mathcal{L} - r)h_b^\xi(x)$ is negative for all $x \in \mathbb{R}$. On the other hand, if there is only one root $\tilde{x}$, then $(\mathcal{L} - r)h_b^\xi(\tilde{x}) = 0$ and $(\mathcal{L} - r)h_b^\xi(x) < 0$ for all other x. In either case, $h_b^\xi(x)/F(x)$ is a strictly decreasing function and (3.37) is true.

Otherwise if $f_b(x) = 0$ has two distinct roots x_{b1} and x_{b2} with $x_{b1} < x_{b2}$, then

$$(\mathcal{L} - r)h_b^\xi(x) \begin{cases} < 0 & \text{if } x \in (-\infty, x_{b1}) \cup (x_{b2}, +\infty), \\ > 0 & \text{if } x \in (x_{b1}, x_{b2}). \end{cases} \tag{3.39}$$

Applying (3.39) to (3.38), the derivative $(h_b^\xi/F)'(x)$ is negative on $(-\infty, x_{b1})$ since the integrand in (3.38) is negative. Hence, $h_b^\xi(x)/F(x)$ is strictly decreasing on $(-\infty, x_{b1})$. We further note that $b^{\xi*} > x_s > x_{b2}$. Observe that on the interval (x_{b1}, x_{b2}), the intergrand is positive. It is therefore possible for $(h_b^\xi/F)'$ to change sign at some $x \in (x_{b1}, x_{b2})$. For this to happen, the positive part of the integral must be larger than the absolute value of the negative part. In other words, (3.23) must hold. If (3.23) holds, then there must exist some $\tilde{a}^* \in (x_{b1}, x_{b2})$ such that $(h_b^\xi/F)'(\tilde{a}^*) = 0$, or equivalently (3.16) holds:

$$\left(\frac{h_b^\xi}{F}\right)'(\tilde{a}^*) = \frac{h_b^{\xi'}(\tilde{a}^*)}{F(\tilde{a}^*)} - \frac{h_b^\xi(\tilde{a}^*)F'(\tilde{a}^*)}{F^2(\tilde{a}^*)} = \frac{e^{\tilde{a}^*}}{F(\tilde{a}^*)} - \frac{(e^{\tilde{a}^*} + c_b)F'(\tilde{a}^*)}{F^2(\tilde{a}^*)}.$$

If (3.16) holds, then we have

$$\left| \int_{-\infty}^{x_{b1}} \Psi(x)e^x f_b(x)dx \right| = \int_{x_{b1}}^{\tilde{a}^*} \Psi(x)e^x f_b(x)dx.$$

In addition, since

$$\int_{\tilde{a}^*}^{x_{b2}} \Psi(x)e^x f_b(x)dx > 0,$$

it follows that

$$\left| \int_{-\infty}^{x_{b1}} \Psi(x)e^x f_b(x)dx \right| < \int_{x_{b1}}^{x_{b2}} \Psi(x)e^x f_b(x)dx.$$

This establishes the equivalence between (3.16) and (3.23). Under this condition, h_b^ξ/F is strictly decreasing on $(x_{b1}, \tilde{a}^*)$. Then, either it is strictly increasing on $(\tilde{a}^*, b^{\xi*})$, or there exists some $\bar{x} \in (x_{b2}, b^{\xi*})$ such that $h_b^\xi(x)/F(x)$ is strictly increasing on $(\tilde{a}^*, \bar{x})$ and strictly decreasing on $(\bar{x}, b^{\xi*})$. In both cases, (3.37) is true if and only if (3.17) holds.

Alternatively, if (3.23) does not hold, then by in (3.38), the integral $(h_b^\xi/F)'$ will always be negative. This means that the function $h_b^\xi(x)/F(x)$ is strictly decreasing for all $x \in (-\infty, b^{\xi*})$, in which case (3.37) holds.

We are thus able to show that (3.34) holds, in particular the minimum of 0 is achieved as a result of (3.36). To prove (3.35), we go through a similar procedure. To check that

$$(r - \mathcal{L})\tilde{V}^\xi(x) \geq 0$$

holds, we consider two cases. First when $x < b^{\xi*}$, we get

$$(r - \mathcal{L})\tilde{V}^\xi(x) = \frac{e^{b^{\xi*}} - c_s}{F(b^{\xi*})}(r - \mathcal{L})F(x) = 0.$$

On the other hand, when $x \geq b^{\xi*}$, the inequality holds

$$(r - \mathcal{L})\tilde{V}^\xi(x) = (r - \mathcal{L})h_s^\xi(x) > 0,$$

since $b^{\xi*} > x_s$ (first inequality of (3.33)).

Similarly, when $x \geq b^{\xi*}$, we have

$$\tilde{V}^\xi(x) - (\tilde{J}^\xi(x) + h_s^\xi(x)) = h_s^\xi(x) - h_s^\xi(x) = 0.$$

When $x < b^{\xi*}$, the inequality holds:

$$\tilde{V}^\xi(x) - (\tilde{J}^\xi(x) + h_s^\xi(x)) = \frac{h_s^\xi(b^{\xi*})}{F(b^{\xi*})}F(x) - h_s^\xi(x) \geq 0,$$

which is equivalent to $\frac{h_s^\xi(x)}{F(x)} \leq \frac{h_s^\xi(b^{\xi*})}{F(b^{\xi*})}$, due to (3.32) and (3.33).

Proof of Theorem 3.7 (Part 1)   Define the functions

$$q_G(x, z) = \int_x^{+\infty} \Phi(s)(\mathcal{L} - r)h_b^\xi(s)ds - \int_z^{+\infty} \Phi(s)(\mathcal{L} - r)h_s^\xi(s)ds,$$

$$q_F(x, z) = \int_{-\infty}^x \Psi(s)(\mathcal{L} - r)h_b^\xi(s)ds - \int_{-\infty}^z \Psi(s)(\mathcal{L} - r)h_s^\xi(s)ds,$$

where Φ and Ψ are given in (3.30) and (3.24), respectively. We look for the points $\tilde{d}^* < \tilde{b}^*$ such that

$$q_G(\tilde{d}^*, \tilde{b}^*) = 0, \quad \text{and} \quad q_F(\tilde{d}^*, \tilde{b}^*) = 0.$$

This is because these two equations are equivalent to (3.21) and (3.22), respectively.

Now we start to solve the equations by first narrowing down the range for $\tilde{d}^*$ and $\tilde{b}^*$. Observe that

$$q_G(x, z) = \int_x^z \Phi(s)(\mathcal{L} - r)h_b^\xi(s)ds + \int_z^\infty \Phi(s)[(\mathcal{L} - r)(h_b^\xi(s) - h_s^\xi(s)]ds$$

$$= \int_x^z \Phi(s)(\mathcal{L} - r)h_b^\xi(s)ds - r(c_b + c_s)\int_z^\infty \Phi(s)ds$$

$$< 0, \tag{3.40}$$

for all x and z such that $x_{b2} \le x < z$. Therefore, $\tilde{d}^* \in (-\infty, x_{b2})$.

Since $b^{\xi*} > x_s$ satisfies $q(b^{\xi*}) = 0$ and $\tilde{a}^* < x_{b2}$ satisfies (3.16), we have

$$\lim_{z \to +\infty} q_F(x, z)$$

$$= \int_{-\infty}^x \Psi(s)(\mathcal{L} - r)h_b^\xi(s)ds - q(b^{\xi*}) - \int_{b^{\xi*}}^{+\infty} \Psi(s)(\mathcal{L} - r)h_s^\xi(s)ds$$

$$> 0,$$

for all $x \in (\tilde{a}^*, x_{b2})$. Also, we note that

$$\frac{\partial q_F}{\partial z}(x, z) = -\Psi(z)(\mathcal{L} - r)h_s^\xi(z) \begin{cases} < 0 & \text{if } z < x_s, \\ > 0 & \text{if } z > x_s, \end{cases} \tag{3.41}$$

and

$$q_F(x, x) = \int_{-\infty}^x \Psi(s)(\mathcal{L} - r)\left[h_b^\xi(s) - h_s^\xi(s)\right]ds$$

$$= -r(c_b + c_s)\int_{-\infty}^x \Psi(s)ds < 0. \tag{3.42}$$

Then, (3.41) and (3.42) imply that there exists a unique function $\beta :$ $[\tilde{a}^*, x_{b2}) \mapsto \mathbb{R}$ s.t. $\beta(x) > x_s$ and

$$q_F(x, \beta(x)) = 0. \tag{3.43}$$

Differentiating (3.43) with respect to x, we see that

$$\beta'(x) = \frac{\Psi(x)(\mathcal{L}-r)h_b^\xi(x)}{\Psi(\beta(x))(\mathcal{L}-r)h_s^\xi(\beta(x))} < 0,$$

for all $x \in (x_{b1}, x_{b2})$. In addition, by the facts that $b^{\xi*} > x_s$ satisfies $q(b^{\xi*}) = 0$, $\tilde{a}^*$ satisfies (3.16), and the definition of q_F, we have

$$\beta(\tilde{a}^*) = b^{\xi*}.$$

By (3.40), we have $\lim_{x\uparrow x_{b2}} q_G(x, \beta(x)) < 0$. By computation, we get that

$$\frac{d}{dx}q_G(x, \beta(x)) = -\frac{\Phi(x)\Psi(\beta(x)) - \Phi(\beta(x))\Psi(x)}{\Psi(\beta(x))}(\mathcal{L}-r)h_b^\xi(x)$$

$$= -\Psi(x)\left[\frac{G(x)}{F(x)} - \frac{G(\beta(x))}{F(\beta(x))}\right](\mathcal{L}-r)h_b^\xi(x) < 0,$$

for all $x \in (x_{b1}, x_{b2})$. Therefore, there exists a unique $\tilde{d}^*$ such that $q_G(\tilde{d}^*, \beta(\tilde{d}^*)) = 0$ if and only if

$$q_G(\tilde{a}^*, \beta(\tilde{a}^*)) > 0.$$

The above inequality holds if (3.18) holds. Indeed, direct computation yields the equivalence:

$$q_G(\tilde{a}^*, \beta(\tilde{a}^*))$$

$$= \int_{\tilde{a}^*}^{+\infty} \Phi(s)(\mathcal{L}-r)h_b^\xi(s)ds - \int_{b^{\xi*}}^{+\infty} \Phi(s)(\mathcal{L}-r)h_s^\xi(s)ds$$

$$= -\frac{h_b^\xi(\tilde{a}^*)}{F(\tilde{a}^*)} - \frac{G(b^{\xi*})}{F(b^{\xi*})}\int_{-\infty}^{b^{\xi*}} \Psi(s)(\mathcal{L}-r)h_s^\xi(s)ds - \int_{b^{\xi*}}^{+\infty} \Phi(s)(\mathcal{L}-r)h_s^\xi(s)ds$$

$$= -\frac{e^{\tilde{a}^*}+c_b}{F(\tilde{a}^*)} + \frac{e^{b^{\xi*}}-c_s}{F(b^{\xi*})}.$$

When this solution exists, we have

$$\tilde{d}^* \in (x_{b1}, x_{b2}) \text{ and } \tilde{b}^* := \beta(\tilde{d}^*) > x_s.$$

Next, we show that the functions $\tilde{J}^\xi$ and $\tilde{V}^\xi$ given in (3.19) and (3.20) satisfy the pair of VIs in (3.34) and (3.35). In the same vein as the proof for the Theorem 3.6, we show

$$(r - \mathcal{L})\tilde{J}^\xi(x) \geq 0$$

by examining the 3 disjoint regions on which $\tilde{J}^\xi(x)$ assume different forms. When $x < \tilde{a}^*$,

$$(r - \mathcal{L})\tilde{J}^\xi(x) = \tilde{P}(r - \mathcal{L})F(x) = 0.$$

Next, when $x > \tilde{d}^*$,
$$(r - \mathcal{L})\tilde{J}^\xi(x) = \tilde{Q}(r - \mathcal{L})G(x) = 0.$$
Finally for $x \in [\tilde{a}^*, \tilde{d}^*]$,
$$(r - \mathcal{L})\tilde{J}^\xi(x) = (r - \mathcal{L})(\tilde{K}F(x) - h_b^\xi(x)) = -(r - \mathcal{L})h_b^\xi(x) > 0,$$
as a result of (3.39) since $\tilde{a}^*, \tilde{d}^* \in (x_{b1}, x_{b2})$.

Next, we verify that
$$(r - \mathcal{L})\tilde{V}^\xi(x) \geq 0.$$
Indeed, we have $(r - \mathcal{L})\tilde{V}^\xi(x) = \tilde{K}(r - \mathcal{L})F(x) = 0$ for $x < \tilde{b}^*$. When $x \geq \tilde{b}^*$, we get the inequality $(r - \mathcal{L})\tilde{V}^\xi(x) = (r - \mathcal{L})(\tilde{Q}G(x) + h_s^\xi(x)) = (r - \mathcal{L})h_s^\xi(x) > 0$ since $\tilde{b}^* > x_s$ and due to (3.14).

It remains to show that $\tilde{J}^\xi(x) - (\tilde{V}^\xi(x) - h_b^\xi(x)) \geq 0$ and $\tilde{V}^\xi(x) - (\tilde{J}^\xi(x) + h_s^\xi(x)) \geq 0$. When $x < \tilde{a}^*$, we have
$$\tilde{J}^\xi(x) - (\tilde{V}^\xi(x) - h_b^\xi(x)) = (\tilde{P} - \tilde{K})F(x) + (e^x + c_b)$$
$$= -F(x)\frac{e^{\tilde{a}^*} + c_b}{F(\tilde{a}^*)} + (e^x + c_b) \geq 0.$$

This inequality holds since we have shown in the proof of Theorem 3.6 that $\frac{h_b^\xi(x)}{F(x)}$ is strictly decreasing for $x < \tilde{a}^*$. In addition,
$$\tilde{V}^\xi(x) - (\tilde{J}^\xi(x) + h_s^\xi(x)) = F(x)\frac{e^{\tilde{a}^*} + c_b}{F(\tilde{a}^*)} - (e^x - c_s) \geq 0,$$
since (3.31) (along with the ensuing explanation) implies that $\frac{h_s^\xi(x)}{F(x)}$ is increasing for all $x \leq \tilde{a}^*$.

In the other region where $x \in [\tilde{a}^*, \tilde{d}^*]$, we have
$$\tilde{J}^\xi(x) - (\tilde{V}^\xi(x) - h_b^\xi(x)) = 0,$$
$$\tilde{V}^\xi(x) - (\tilde{J}^\xi(x) + h_s^\xi(x)) = h_b^\xi(x) - h_s^\xi(x) = c_b + c_s \geq 0.$$
When $x > \tilde{b}^*$, it is clear that
$$\tilde{J}^\xi(x) - (\tilde{V}^\xi(x) - h_b^\xi(x)) = h_b^\xi(x) - h_s^\xi(x) = c_b + c_s \geq 0,$$
$$\tilde{V}^\xi(x) - (\tilde{J}^\xi(x) + h_s^\xi(x)) = 0.$$
To establish the inequalities for $x \in (\tilde{d}^*, \tilde{b}^*)$, we first denote
$$g_{\tilde{J}^\xi}(x) := \tilde{J}^\xi(x) - (\tilde{V}^\xi(x) - h_b^\xi(x)) = \tilde{Q}G(x) - \tilde{K}F(x) + h_b^\xi(x)$$
$$= F(x)\int_{\tilde{d}^*}^x \Phi(s)(\mathcal{L} - r)h_b^\xi(s)ds - G(x)\int_{\tilde{d}^*}^x \Psi(s)(\mathcal{L} - r)h_b^\xi(s)ds,$$
$$g_{\tilde{V}^\xi}(x) := \tilde{V}^\xi(x) - (\tilde{J}^\xi(x) + h_s^\xi(x)) = \tilde{K}F(x) - \tilde{Q}G(x) - h_s^\xi(x)$$
$$= F(x)\int_x^{\tilde{b}^*} \Phi(s)(\mathcal{L} - r)h_s^\xi(s)ds - G(x)\int_x^{\tilde{b}^*} \Psi(s)(\mathcal{L} - r)h_s^\xi(s)ds.$$

In turn, we compute to get

$$g'_{\tilde{J}^\xi}(x) = F'(x) \int_{\tilde{d}^*}^x \Phi(s)(\mathcal{L}-r)h_b^\xi(s)ds - G'(x) \int_{\tilde{d}^*}^x \Psi(s)(\mathcal{L}-r)h_b^\xi(s)ds,$$

$$g'_{\tilde{V}^\xi}(x) = F'(x) \int_x^{\tilde{b}^*} \Phi(s)(\mathcal{L}-r)h_s^\xi(s)ds - G'(x) \int_x^{\tilde{b}^*} \Psi(s)(\mathcal{L}-r)h_s^\xi(s)ds.$$

Recall the definition of x_{b2} and x_s, and the fact that $G' < 0 < F'$, we have $g'_{\tilde{J}^\xi}(x) > 0$ for $x \in (\tilde{d}^*, x_{b2})$ and $g'_{\tilde{V}^\xi}(x) < 0$ for $x \in (x_s, \tilde{b}^*)$. These, together with the fact that $g_{\tilde{J}^\xi}(\tilde{d}^*) = g_{\tilde{V}^\xi}(\tilde{b}^*) = 0$, imply that

$$g_{\tilde{J}^\xi}(x) > 0 \text{ for } x \in (\tilde{d}^*, x_{b2}), \text{ and } g_{\tilde{V}^\xi}(x) > 0 \text{ for } x \in (x_s, \tilde{b}^*).$$

Furthermore, since we have

$$g_{\tilde{J}^\xi}(\tilde{b}^*) = c_b + c_s \geq 0, \quad g_{\tilde{V}^\xi}(\tilde{d}^*) = c_b + c_s \geq 0, \tag{3.44}$$

and

$$(\mathcal{L}-r)g_{\tilde{J}^\xi}(x) = (\mathcal{L}-r)h_b^\xi(x) < 0 \text{ for all } x \in (x_{b2}, \tilde{b}^*),$$

$$(\mathcal{L}-r)g_{\tilde{V}^\xi}(x) = -(\mathcal{L}-r)h_s^\xi(x) < 0 \text{ for all } x \in (\tilde{d}^*, x_s). \tag{3.45}$$

In view of inequalities (3.44)–(3.45), the maximum principle implies that $g_{\tilde{J}^\xi}(x) \geq 0$ and $g_{\tilde{V}^\xi}(x) \geq 0$ for all $x \in (\tilde{d}^*, \tilde{b}^*)$. Hence, we conclude that $\tilde{J}(x) - (\tilde{V}(x) - h_b^\xi(x)) \geq 0$ and $\tilde{V}(x) - (\tilde{J}(x) + h_s^\xi(x)) \geq 0$ hold for $x \in (\tilde{d}^*, \tilde{b}^*)$.

Proof of Theorems 3.6 and 3.7 (Part 2) We now show that the candidate solutions in Theorems 3.6 and 3.7, denoted by $\tilde{j}^\xi$ and $\tilde{v}^\xi$, are equal to the optimal switching value functions $\tilde{J}^\xi$ and $\tilde{V}^\xi$ in (3.6) and (3.7), respectively. First, we note that $\tilde{j}^\xi \leq \tilde{J}^\xi$ and $\tilde{v}^\xi \leq \tilde{V}^\xi$, since $\tilde{J}^\xi$ and $\tilde{V}^\xi$ dominate the expected discounted cash flow from any admissible strategy.

Next, we show the reverse inequaities. In Part 1, we have proved that $\tilde{j}^\xi$ and $\tilde{v}^\xi$ satisfy the VIs (3.34) and (3.35). In particular, we know that $(r - \mathcal{L})\tilde{j}^\xi \geq 0$, and $(r - \mathcal{L})\tilde{v}^\xi \geq 0$. Then by Dynkin's formula and Fatou's lemma, as in Øksendal (2003, p. 226), for any stopping times ζ_1 and ζ_2 such that $0 \leq \zeta_1 \leq \zeta_2$ almost surely, we have the inequalities

$$\mathbb{E}_x\{e^{-r\zeta_1}\tilde{j}^\xi(X_{\zeta_1})\} \geq \mathbb{E}_x\{e^{-r\zeta_2}\tilde{j}^\xi(X_{\zeta_2})\}, \tag{3.46}$$

$$\mathbb{E}_x\{e^{-r\zeta_1}\tilde{v}^\xi(X_{\zeta_1})\} \geq \mathbb{E}_x\{e^{-r\zeta_2}\tilde{v}^\xi(X_{\zeta_2})\}. \tag{3.47}$$

For $\Lambda_0 = (\nu_1, \tau_1, \nu_2, \tau_2, \dots)$, noting that $\nu_1 \leq \tau_1$ almost surely, we have

$$\tilde{j}^\xi(x) \geq \mathbb{E}_x\{e^{-r\nu_1}\tilde{j}^\xi(X_{\nu_1})\} \tag{3.48}$$

$$\geq \mathbb{E}_x\{e^{-r\nu_1}(\tilde{v}^\xi(X_{\nu_1}) - h_b^\xi(X_{\nu_1}))\} \tag{3.49}$$

$$\geq \mathbb{E}_x\{e^{-r\tau_1}\tilde{v}^\xi(X_{\tau_1})\} - \mathbb{E}_x\{e^{-r\nu_1}h_b^\xi(X_{\nu_1})\} \tag{3.50}$$

$$\geq \mathbb{E}_x\{e^{-r\tau_1}(\tilde{j}^\xi(X_{\tau_1}) + h_s^\xi(X_{\tau_1}))\} - \mathbb{E}_x\{e^{-r\nu_1}h_b^\xi(X_{\nu_1})\} \tag{3.51}$$

$$= \mathbb{E}_x\{e^{-r\tau_1}\tilde{j}^\xi(X_{\tau_1})\} + \mathbb{E}_x\{e^{-r\tau_1}h_s^\xi(X_{\tau_1}) - e^{-r\nu_1}h_b^\xi(X_{\nu_1})\}, \tag{3.52}$$

where (3.48) and (3.50) follow from (3.46) and (3.47) respectively. Also, (3.49) and (3.51) follow from (3.34) and (3.35) respectively. Observe that (3.52) is a recursion and $\tilde{j}^\xi(x) \geq 0$ in both Theorems 3.6 and 3.7, we obtain

$$\tilde{j}^\xi(x) \geq \mathbb{E}_x\left\{\sum_{n=1}^\infty [e^{-r\tau_n}h_s^\xi(X_{\tau_n}) - e^{-r\nu_n}h_b^\xi(X_{\nu_n})]\right\}.$$

Maximizing over all Λ_0 yields that $\tilde{j}^\xi(x) \geq \tilde{J}^\xi(x)$. A similar proof gives $\tilde{v}^\xi(x) \geq \tilde{V}^\xi(x)$.

Remark 3.10. If there is no transaction cost for entry, i.e. $c_b = 0$, then f_b, which is now a linear function with a non-zero slope, has one root x_0. Moreover, we have $f_b(x) > 0$ for $x \in (-\infty, x_0)$ and $f_b(x) < 0$ for $x \in (x_0, +\infty)$. This implies that the entry region must be of the form $(-\infty, d_0)$, for some number d_0. Hence, the continuation region for entry is the *connected* interval (d_0, ∞).

Remark 3.11. Let $\mathcal{L}^\xi$ be the infinitesimal generator of the XOU process $\xi = e^X$, and define the function $H_b(\varsigma) := \varsigma + c_b \equiv h_b^\xi(\ln \varsigma)$. In other words, we have the equivalence:

$$(\mathcal{L}^\xi - r)H_b(\varsigma) \equiv (\mathcal{L} - r)h_b^\xi(\ln \varsigma).$$

Referring to (3.13) and (3.14), we have *either* that

$$(\mathcal{L}^\xi - r)H_b(\varsigma) \begin{cases} > 0 & \text{for } \varsigma \in (\varsigma_{b1}, \varsigma_{b2}), \\ < 0 & \text{for } \varsigma \in (0, \varsigma_{b1}) \cup (\varsigma_{b2}, +\infty), \end{cases} \tag{3.53}$$

where $\varsigma_{b1} = e^{x_{b1}} > 0$ and $\varsigma_{b2} = e^{x_{b2}}$ and $x_{b1} < x_{b2}$ are two distinct roots to (3.13), *or*

$$(\mathcal{L}^\xi - r)H_b(\varsigma) < 0, \quad \text{for } \varsigma \in (0, \varsigma^*) \cup (\varsigma^*, +\infty), \tag{3.54}$$

where $\varsigma^* = e^{x_b}$ and x_b is the single root to (3.13). In both cases, Assumption 4 of Zervos *et al.* (2013) is violated, and their results cannot be applied.

Indeed, they would require that $(\mathcal{L}^\xi - r)H_b(\varsigma)$ is strictly negative over a connected interval of the form (ς_0, ∞), for some fixed $\varsigma_0 \geq 0$. However, it is clear from (3.53) and (3.54) that such a region is disconnected.

In fact, the approach by Zervos *et al.* (2013) applies to the optimal switching problems where the optimal wait-for-entry region (in log-price) is of the form $(\tilde{d}^*, \infty)$, rather than the *disconnected* region $(-\infty, \tilde{a}^*) \cup (\tilde{d}^*, \infty)$, as in our case with an XOU underlying. Using the new inferred structure of the wait-for-entry region, we have modified the arguments in Zervos *et al.* (2013) to solve our optimal switching problem for Theorems 3.6 and 3.7.

3.4 Proofs of Lemmas

We now present detailed proofs on the properties of V^ξ, J^ξ, H^ξ and $\hat{H}^\xi$.

Proof of Lemma 3.1 (Bounds of V^ξ)

First, by Dynkin's formula, we have every $x \in \mathbb{R}$ and $\tau \in \mathcal{T}$,

$$\mathbb{E}_x\{e^{-r\tau}e^{X_\tau}\} - e^x = \mathbb{E}_x\left\{\int_0^\tau e^{-rt}(\mathcal{L} - r)e^{X_t}dt\right\}$$

$$= \mathbb{E}_x\left\{\int_0^\tau e^{-rt}e^{X_t}\left(\frac{\sigma^2}{2} + \mu\theta - r - \mu X_t\right)dt\right\}.$$

The function $e^x\left(\frac{\sigma^2}{2} + \mu\theta - r - \mu x\right)$ is bounded above on $\mathbb{R}$. Let M be an upper bound, it follows that

$$\mathbb{E}_x\left\{e^{-r\tau}e^{X_\tau}\right\} - e^x \leq M\mathbb{E}\left\{\int_0^\tau e^{-rt}dt\right\}$$

$$\leq M\mathbb{E}\left\{\int_0^{+\infty} e^{-rt}dt\right\} = \frac{M}{r} := K^\xi.$$

Since $h_s^\xi(x) = e^x - c_s \leq e^x$, we have

$$\mathbb{E}_x\left\{e^{-r\tau}h_s^\xi(X_\tau)\right\} \leq \mathbb{E}_x\left\{e^{-r\tau}e^{X_\tau}\right\} \leq e^x + K^\xi.$$

Therefore, $V^\xi(x) \leq e^x + K^\xi$. Lastly, the choice of $\tau = +\infty$ as a candidate stopping time implies that $V^\xi(x) \geq 0$.

Proof of Lemma 3.3 (Bounds of J^ξ)

From the limit

$$\limsup_{x\to-\infty} (\hat{h}^\xi(x))^+ = \limsup_{x\to-\infty} (V^\xi(x) - e^x - c_b)^+ = 0,$$

it follows that there exists some $\hat{x}_0^\xi$ such that $(\hat{h}^\xi(x))^+ \leq \hat{K}_1$ for every $x \in (-\infty, \hat{x}_0^\xi)$ and some positive constant $\hat{K}_1$. Next, $(\hat{h}^\xi(x))^+$ is bounded by some positive constant $\hat{K}_2$ on the closed interval $[\hat{x}_0^\xi, b^{\xi*}]$. Also, $(\hat{h}^\xi(x))^+ = (V^\xi(x) - e^x - c_b)^+ = (-(c_s + c_b))^+ = 0$ for $x \geq b^{\xi*}$. Taking $\hat{K}^\xi = \hat{K}_1 \vee \hat{K}_2$, we have $(\hat{h}^\xi(x))^+ \leq \hat{K}^\xi$ for all $x \in \mathbb{R}$. This yields the inequality

$$\mathbb{E}_x\{e^{-r\tau}\hat{h}^\xi(X_\tau)\} \leq \mathbb{E}_x\{e^{-r\tau}(\hat{h}^\xi(X_\tau))^+\} \leq \mathbb{E}_x\{e^{-r\tau}\hat{K}^\xi\} \leq \hat{K}^\xi,$$

for every $x \in \mathbb{R}$ and every $\tau \in \mathcal{T}$. Hence, $J^\xi(x) \leq \hat{K}^\xi$. The admissibility of $\tau = +\infty$ yields $J^\xi(x) \geq 0$.

Proof of Lemma 3.5 (Bounds of $\tilde{J}^\xi$ and $\tilde{V}^\xi$)

By definition, both $\tilde{J}^\xi(x)$ and $\tilde{V}^\xi(x)$ are nonnegative. Using Dynkin's formula, we have

$$\mathbb{E}_x\left\{e^{-r\tau_n}e^{X_{\tau_n}}\right\} - \mathbb{E}_x\left\{e^{-r\nu_n}e^{X_{\nu_n}}\right\}$$

$$= \mathbb{E}_x\left\{\int_{\nu_n}^{\tau_n} e^{-rt}(\mathcal{L}-r)e^{X_t}dt\right\}$$

$$= \mathbb{E}_x\left\{\int_{\nu_n}^{\tau_n} e^{-rt}e^{X_t}\left(\frac{\sigma^2}{2} + \mu\theta - r - \mu X_t\right)dt\right\}.$$

As we have pointed out in Section 3.4, the function $e^x\left(\frac{\sigma^2}{2} + \mu\theta - r - \mu x\right)$ is bounded above on $\mathbb{R}$ and M is an upper bound. It follows that

$$\mathbb{E}_x\left\{e^{-r\tau_n}e^{X_{\tau_n}}\right\} - \mathbb{E}_x\left\{e^{-r\nu_n}e^{X_{\nu_n}}\right\} \leq M\mathbb{E}_x\left\{\int_{\nu_n}^{\tau_n} e^{-rt}dt\right\}.$$

Since $e^x - c_s \leq e^x$ and $e^x + c_b \geq e^x$, we have

$$\mathbb{E}_x\left\{\sum_{n=1}^\infty [e^{-r\tau_n}h_s^\xi(X_{\tau_n}) - e^{-r\nu_n}h_b^\xi(X_{\nu_n})]\right\}$$

$$\leq \sum_{n=1}^\infty \left(\mathbb{E}_x\left\{e^{-r\tau_n}e^{X_{\tau_n}}\right\} - \mathbb{E}_x\left\{e^{-r\nu_n}e^{X_{\nu_n}}\right\}\right)$$

$$\leq \sum_{n=1}^\infty M\mathbb{E}_x\left\{\int_{\nu_n}^{\tau_n} e^{-rt}dt\right\} \leq M\int_0^\infty e^{-rt}dt = \frac{M}{r} := C_1,$$

which implies that $0 \leq \tilde{J}^\xi(x) \leq C_1$. Similarly,

$$\mathbb{E}_x\left\{e^{-r\tau_1}h_s^\xi(X_{\tau_1}) + \sum_{n=2}^\infty [e^{-r\tau_n}h_s^\xi(X_{\tau_n}) - e^{-r\tau_n}h_b^\xi(X_{\tau_n})]\right\}$$

$$\leq C_1 + \mathbb{E}_x\left\{e^{-r\tau_1}h_b^\xi(X_{\tau_1})\right\}.$$

Letting $\nu_1 = 0$ and using Dynkin's formula again, we have

$$\mathbb{E}_x \left\{ e^{-r\tau_1} e^{X_{\tau_1}} \right\} - e^x \leq \frac{M}{r}.$$

This implies that

$$\tilde{V}^\xi(x) \leq C_1 + e^x + \frac{M}{r} := e^x + C_2.$$

Proof of Lemma 3.8 (Properties of H^ξ)

The continuity and twice differentiability of H^ξ on $(0, +\infty)$ follow directly from those of h_s^ξ, G and ψ. On the other hand, we have $H^\xi(0) := \lim_{x \to -\infty} \frac{(h_s^\xi(x))^+}{G(x)} = \lim_{x \to -\infty} \frac{(e^x - c_s)^+}{G(x)} = \lim_{x \to -\infty} \frac{0}{G(x)} = 0$. Hence, the continuity of H^ξ at 0 follows from

$$\lim_{z \to 0} H^\xi(z) = \lim_{x \to -\infty} \frac{h_s^\xi(x)}{G(x)} = \lim_{x \to -\infty} \frac{e^x - c_s}{G(x)} = 0.$$

Next, we prove properties (i)-(iii) of H^ξ.

(i) This follows trivially from the fact that $\psi(x)$ is a strictly increasing function and $G(x) > 0$.

(ii) By the definition of H^ξ,

$$H^{\xi'}(z) = \frac{1}{\psi'(x)} \left(\frac{h_s^\xi}{G} \right)'(x) = \frac{[e^x G(x) - (e^x - c_s)G'(x)]}{\psi'(x)G^2(x)}, \quad z = \psi(x).$$

For $x \in (\ln c_s, +\infty)$, $e^x - c_s > 0$, $G'(x) < 0$, so $e^x G(x) - (e^x - c_s)G'(x) > 0$. Also, since both $\psi'(x)$ and $G^2(x)$ are positive, we conclude that $H^{\xi'}(z) > 0$ for $z \in (\psi(\ln c_s), +\infty)$.

The proof of the limit of $H^{\xi'}(z)$ will make use of property (iii), and is thus deferred until after the proof of property (iii).

(iii) By differentiation, we have

$$H^{\xi''}(z) = \frac{2}{\sigma^2 G(x)(\psi'(x))^2} [(\mathcal{L} - r)h_s^\xi](x), \quad z = \psi(x).$$

Since σ^2, $G(x)$ and $(\psi'(x))^2$ are all positive, we only need to determine the sign of $(\mathcal{L} - r)h_s^\xi(x) = e^x f_s(x)$. Hence, property (iii) follows from (3.14).

To find the limit of $H^{\xi'}(z)$, we first observe that

$$\lim_{x \to +\infty} \frac{h_s^\xi(x)}{F(x)} = 0. \tag{3.55}$$

Indeed, we have

$$
\lim_{x \to +\infty} \frac{h_s^\xi(x)}{F(x)} = \lim_{x \to +\infty} \frac{1}{e^{-x} F(x)}
$$

$$
= \lim_{x \to +\infty} \left(\int_0^{+\infty} u^{\frac{r}{\mu}-1} e^{(\sqrt{\frac{2\mu}{\sigma^2}} - \frac{1}{u})xu - \sqrt{\frac{2\mu}{\sigma^2}}\theta u - \frac{u^2}{2}} \, du \right)^{-1}
$$

$$
= \lim_{x \to +\infty} \left(\int_0^{\sqrt{\frac{\sigma^2}{2\mu}}} u^{\frac{r}{\mu}-1} e^{(\sqrt{\frac{2\mu}{\sigma^2}} - \frac{1}{u})xu - \sqrt{\frac{2\mu}{\sigma^2}}\theta u - \frac{u^2}{2}} \, du \right.
$$

$$
\left. + \int_{\sqrt{\frac{\sigma^2}{2\mu}}}^{+\infty} u^{\frac{r}{\mu}-1} e^{(\sqrt{\frac{2\mu}{\sigma^2}} - \frac{1}{u})xu - \sqrt{\frac{2\mu}{\sigma^2}}\theta u - \frac{u^2}{2}} \, du \right)^{-1}.
$$

Since the first term on the RHS is non-negative and the second term is strictly increasing and convex in x, the limit is zero.

Turning now to $H^{\xi'}(z)$, we note that

$$
H^{\xi'}(z) = \frac{1}{\psi'(x)} \left(\frac{h_s^\xi}{G} \right)'(x), \quad z = \psi(x).
$$

As we have shown, for $z > \psi(\ln c_s) \wedge \psi(x_s)$, $H^{\xi'}(z)$ is a positive and decreasing function. Hence the limit exists and satisfies

$$
\lim_{z \to +\infty} H^{\xi'}(z) = \lim_{x \to +\infty} \frac{1}{\psi'(x)} \left(\frac{h_s^\xi}{G} \right)'(x) = c \geq 0. \tag{3.56}
$$

Observe that $\lim_{x \to +\infty} \frac{h_s^\xi(x)}{G(x)} = +\infty$, $\lim_{x \to +\infty} \psi(x) = +\infty$, and $\lim_{x \to +\infty} \frac{\left(\frac{h_s^\xi(x)}{G(x)} \right)'}{\psi'(x)}$ exists, and $\psi'(x) \neq 0$. We can apply L'Hopital's rule to get

$$
\lim_{x \to +\infty} \frac{h_s^\xi(x)}{F(x)} = \lim_{x \to +\infty} \frac{\frac{h_s^\xi(x)}{G(x)}}{\frac{F(x)}{G(x)}} = \lim_{x \to +\infty} \frac{\left(\frac{h_s^\xi(x)}{G(x)} \right)'}{\psi'(x)} = c. \tag{3.57}
$$

Comparing (3.55) and (3.57) implies that $c = 0$. From (3.56), we conclude that $\lim_{z \to +\infty} H^{\xi'}(z) = 0$.

Proof of Lemma 3.9 (Properties of $\hat{H}^\xi$)

It is straightforward to check that $V^\xi(x)$ is continuous and differentiable everywhere, and twice differentiable everywhere except at $x = b^{\xi*}$. The same properties hold for $\hat{h}^\xi(x)$. Since both G and ψ are twice differentiable everywhere, the continuity and differentiability of $\hat{H}^\xi$ on $(0, +\infty)$ and twice differentiability on $(0, \psi(b^{\xi*})) \cup (\psi(b^{\xi*}), +\infty)$ follow directly.

To see the continuity of $\hat{H}^\xi(z)$ at 0, note that $V^\xi(x) \to 0$ and $e^x \to 0$ as $x \to -\infty$. Then we have

$$\hat{H}^\xi(0) := \lim_{x \to -\infty} \frac{(\hat{h}^\xi(x))^+}{G(x)} = \lim_{x \to -\infty} \frac{(V^\xi(x) - e^x - c_b)^+}{G(x)} = \lim_{x \to -\infty} \frac{0}{G(x)} = 0,$$

and $\lim_{z \to 0} \hat{H}^\xi(z) = \lim_{x \to -\infty} \frac{\hat{h}^\xi(x)}{G(x)} = \lim_{x \to -\infty} \frac{-c_b}{G(x)} = 0$. There follows the continuity at 0.

(i) For $x \in [b^{\xi*}, +\infty)$, we have $\hat{h}^\xi(x) \equiv -(c_s + c_b) < 0$. Next, the limits $\lim_{x \to -\infty} V^\xi(x) \to 0$ and $\lim_{x \to -\infty} e^x \to 0$ imply that $\lim_{x \to -\infty} \hat{h}^\xi(x) = V^\xi(x) - e^x - c_b \to -c_b < 0$. Therefore, there exists some $\underline{b}^\xi$ such that $\hat{h}^\xi(x) < 0$ for $x \in (-\infty, \underline{b}^\xi)$. For the non-trivial case in question, $\hat{h}^\xi(x)$ must be positive for some x, so we must have $\underline{b}^\xi < b^{\xi*}$. To conclude, we have $\hat{h}^\xi(x) < 0$ for $x \in (-\infty, \underline{b}^\xi) \cup [b^{\xi*}, +\infty)$. This, along with the facts that $\psi(x) \in (0, +\infty)$ is a strictly increasing function and $G(x) > 0$, implies property (i).

(ii) By differentiating $\hat{H}^\xi(z)$, we get

$$\hat{H}^{\xi'}(z) = \frac{1}{\psi'(x)} \left(\frac{\hat{h}^\xi}{G}\right)'(x), \quad z = \psi(x).$$

To determine the sign of $\hat{H}^{\xi'}$, we observe that, for $x \geq b^{\xi*}$,

$$\left(\frac{\hat{h}^\xi(x)}{G(x)}\right)' = \left(\frac{-(c_s + c_b)}{G(x)}\right)' = \frac{(c_s + c_b)G'(x)}{G^2(x)} < 0.$$

Also, $\psi'(x) > 0$ for $x \in \mathbb{R}$. Therefore, $\hat{H}^\xi(z)$ is strictly decreasing for $z \geq \psi(b^{\xi*})$.

(iii) To study the convexity/concavity, we look at the second derivative

$$\hat{H}^{\xi''}(z) = \frac{2}{\sigma^2 G(x)(\psi'(x))^2}(\mathcal{L} - r)\hat{h}^\xi(x), \quad z = \psi(x).$$

Since $\sigma^2, G(x)$ and $(\psi'(x))^2$ are all positive, we only need to determine the sign of $(\mathcal{L} - r)\hat{h}^\xi(x)$:

$$(\mathcal{L} - r)\hat{h}^\xi(x)$$
$$= \frac{\sigma^2}{2}(V^{\xi''}(x) - e^x) + \mu(\theta - x)(V^{\xi'}(x) - e^x) - r(V^\xi(x) - e^x - c_b)$$
$$= \begin{cases} [\mu x - (\mu\theta + \frac{\sigma^2}{2} - r)]e^x + rc_b & \text{if } x \in (-\infty, b^{\xi*}), \\ r(c_s + c_b) > 0 & \text{if } x \in (b^{\xi*}, +\infty). \end{cases}$$

which suggests that $\hat{H}^\xi(z)$ is convex for $z \in (\psi(b^{\xi*}), +\infty)$.

Furthermore, for $x \in (x_s, b^{\xi *})$, we have

$$(\mathcal{L} - r)\hat{h}^\xi(x) = [\mu x - (\mu\theta + \frac{\sigma^2}{2} - r)]e^x + rc_b$$
$$= -e^x f_s(x) + r(c_s + c_b) > r(c_s + c_b) > 0,$$

by the definition of x_s. Therefore, $\hat{H}^\xi(z)$ is also convex on $(\psi(x_s), \psi(b^{\xi *}))$. Thus far, we have established that $\hat{H}^\xi(z)$ is convex on $(\psi(x_s), +\infty)$.

Next, we determine the convexity of $\hat{H}^\xi(z)$ on $(0, \psi(x_s)]$. Denote $\hat{z}_1^\xi :=$ arg max$_{z \in [0,+\infty)} \hat{H}^\xi(z)$. Since sup$_{x \in \mathbb{R}} \hat{h}^\xi(x) > 0$, we must have

$$\hat{H}^\xi(\hat{z}_1^\xi) = \sup_{z \in [0,+\infty)} \hat{H}^\xi(z) > 0.$$

By its continuity and differentiability, $\hat{H}^\xi$ must be concave at $\hat{z}_1^\xi$. Then, there must exist some interval $(\psi(a^{(0)}), \psi(d^{(0)}))$ over which $\hat{H}^\xi$ is concave and $\hat{z}_1^\xi \in (\psi(a^{(0)}), \psi(d^{(0)}))$.

On the other hand, for $x \in (-\infty, x_s]$,

$$((\mathcal{L} - r)\hat{h}^\xi)'(x) = [\mu x - (\mu\theta + \frac{\sigma^2}{2} - r - \mu)]e^x \begin{cases} < 0 & \text{if } x \in (-\infty, x^{\xi *}), \\ > 0 & \text{if } x \in (x^{\xi *}, x_s], \end{cases}$$

where $x^{\xi *} = \theta + \frac{\sigma^2}{2\mu} - \frac{r}{\mu} - 1$. Therefore, $(\mathcal{L} - r)\hat{h}^\xi(x)$ is strictly decreasing on $(-\infty, x^{\xi *})$, strictly increasing on $(x^{\xi *}, x_s]$, and is strictly positive at x_s and $-\infty$:

$$(\mathcal{L} - r)\hat{h}^\xi(x_s) = r(c_s + c_b) > 0 \quad \text{and} \quad \lim_{x \to -\infty} (\mathcal{L} - r)\hat{h}^\xi(x) = rc_b > 0.$$

If $(\mathcal{L} - r)\hat{h}^\xi(x^{\xi *}) = -\mu e^{x^{\xi *}} + rc_b < 0$, then there exist exactly two distinct roots to the equation $(\mathcal{L} - r)\hat{h}^\xi(x) = 0$, denoted as x_{b1} and x_{b2}, such that $-\infty < x_{b1} < x^{\xi *} < x_{b2} < x_s$ and

$$(\mathcal{L} - r)\hat{h}^\xi(x) \begin{cases} > 0 & \text{if } x \in (-\infty, x_{b1}) \cup (x_{b2}, x_s], \\ < 0 & \text{if } x \in (x_{b1}, x_{b2}). \end{cases}$$

On the other hand, if $(\mathcal{L}-r)\hat{h}^\xi(x^{\xi *}) = -\mu e^{x^{\xi *}} + rc_b \geq 0$, then $(\mathcal{L}-r)\hat{h}^\xi(x) \geq 0$ for all $x \in \mathbb{R}$, and $\hat{H}^\xi(z)$ is convex for all z, which contradicts with the existence of a concave interval. Hence, we conclude that $-\mu e^{x^{\xi *}} + rc_b < 0$, and (x_{b1}, x_{b2}) is the unique interval that $(\mathcal{L} - r)\hat{h}^\xi(x) < 0$. Consequently, $(a^{(0)}, d^{(0)})$ coincides with (x_{b1}, x_{b2}) and $\hat{z}_1^\xi \in (\psi(x_{b1}), \psi(x_{b2}))$. This completes the proof.

Chapter 4

Trading Under the CIR Model

In this chapter, we study the problem of trading under the CIR model. We formulate an optimal double stopping problem and an optimal switching problem, and rigorously prove that the optimal starting and stopping strategies are of threshold type.

A CIR process $(Y_t)_{t\geq 0}$ satisfies the SDE

$$dY_t = \mu(\theta - Y_t)\,dt + \sigma\sqrt{Y_t}\,dB_t, \tag{4.1}$$

with constants $\mu, \theta, \sigma > 0$. If $2\mu\theta \geq \sigma^2$ holds, which is often referred to as the Feller condition (see Feller (1951)), then the level 0 is inaccessible by Y. If the initial value $Y_0 > 0$, then Y stays strictly positive at all times almost surely. Nevertheless, if $Y_0 = 0$, then Y will enter the interior of the state space immediately and stays positive thereafter almost surely. If $2\mu\theta < \sigma^2$, then the level 0 is a reflecting boundary. This means that once Y reaches 0, it immediately returns to the interior of the state space and continues to evolve. For a detailed categorization of boundaries for diffusion processes, we refer to Chapter 2 of Borodin and Salminen (2002) and Chapter 15 of Karlin and Taylor (1981).

The CIR conditional probability density of Y_{t_i} at time t_i given $Y_{t_{i-1}} = y_{i-1}$ with time increment $\Delta t = t_i - t_{i-1}$ is given by

$$f^{CIR}(y_i|y_{i-1}; \theta, \mu, \sigma)$$

$$= \frac{1}{\tilde{\sigma}^2}\exp\left(-\frac{y_i + y_{i-1}e^{-\mu\Delta t}}{\tilde{\sigma}^2}\right)\left(\frac{y_i}{y_{i-1}e^{-\mu\Delta t}}\right)^{\frac{q}{2}}I_q\left(\frac{2}{\tilde{\sigma}^2}\sqrt{y_iy_{i-1}e^{-\mu\Delta t}}\right),$$

with the constants

$$\tilde{\sigma}^2 = \sigma^2\frac{1 - e^{-\mu\Delta t}}{2\mu}, \quad q = \frac{2\mu\theta}{\sigma^2} - 1,$$

and $I_q(z)$ is modified Bessel function of the first kind and of order q. See Cox *et al.* (1985).

Using the observed values $(y_i)_{i=0,1,\ldots,n}$, the CIR model parameters can be estimated by maximizing the average log-likelihood:

$$\ell(\theta, \mu, \sigma | y_0, y_1, \ldots, y_n) := \frac{1}{n} \sum_{i=1}^{n} \ln f^{CIR}(y_i | y_{i-1}; \theta, \mu, \sigma)$$

$$= -2 \ln(\tilde{\sigma}) - \frac{1}{n\tilde{\sigma}^2} \sum_{i=1}^{n} (y_i + y_{i-1} e^{-\mu \Delta t})$$

$$+ \frac{1}{n} \sum_{i=1}^{n} \left(\frac{q}{2} \ln \left(\frac{y_i}{y_{i-1} e^{-\mu \Delta t}} \right) + \ln I_q \left(\frac{2}{\tilde{\sigma}^2} \sqrt{y_i y_{i-1} e^{-\mu \Delta t}} \right) \right).$$

For more details on the implementation of the maximum likelihood estimation (MLE) for the CIR process, we refer to Kladivko (2007).

In Section 4.1, we formulate both the optimal starting-stopping and optimal switching problems. Then, we present our analytical results and numerical examples in Section 4.2. The proofs of our main results are detailed in Section 4.3.

4.1 Optimal Trading Problems

Denote by $\mathbb{F}$ the filtration generated by B, and $\mathcal{T}$ the set of all $\mathbb{F}$-stopping times. If a decision to sell is made at some time $\tau \in \mathcal{T}$, then the amount Y_τ is received and simultaneously the constant transaction cost $c_s > 0$ has to be paid. On the other hand, at the time of market entry, a constant transaction cost $c_b > 0$ is incurred.

4.1.1 *Optimal Starting-Stopping Approach*

Given a CIR process, we first consider the optimal timing to stop. The maximum expected discounted value is obtained by solving the optimal stopping problem

$$V^\chi(y) = \sup_{\tau \in \mathcal{T}} \mathbb{E}_y \left\{ e^{-r\tau} (Y_\tau - c_s) \right\}, \tag{4.2}$$

where $r > 0$ is the constant discount rate, and $\mathbb{E}_y\{\cdot\} \equiv \mathbb{E}\{\cdot | Y_0 = y\}$.

The value function V^χ represents the expected value from optimally stopping the process Y. On the other hand, the process value plus the transaction cost constitute the total cost to start. Before even starting, one needs to choose the optimal timing to start, or not to start at all. This leads us to analyze the starting timing inherent in the starting-stopping (or double stopping) problem. Precisely, we solve

$$J^{\chi}(y) = \sup_{\nu \in \mathcal{T}} \mathbb{E}_y \left\{ e^{-r\nu} (V^{\chi}(Y_\nu) - Y_\nu - c_b) \right\}, \qquad (4.3)$$

with the transaction cost c_b incurred upon market entry. In other words, the objective is to maximize the expected difference between the value function $V^{\chi}(Y_\nu)$ and the current Y_ν, minus transaction cost c_b. The value function $J^{\chi}(y)$ represents the maximum expected value that can be gained by entering and subsequently exiting, with transaction costs c_b and c_s incurred, respectively, on entry and exit. For our analysis, the transaction costs c_b and c_s can be different. To facilitate presentation, we denote the functions

$$h_s(y) = y - c_s, \quad \text{and} \quad h_b(y) = y + c_b. \qquad (4.4)$$

If it turns out that $J^{\chi}(Y_0) \leq 0$ for some initial value Y_0, then it is optimal not to start at all. Therefore, it is important to identify the trivial cases. Under the CIR model, since $\sup_{y \in \mathbb{R}_+} (V^{\chi}(y) - h_b(y)) \leq 0$ implies that $J^{\chi}(y) = 0$ for $y \in \mathbb{R}_+$, we shall therefore focus on the case with

$$\sup_{y \in \mathbb{R}_+} (V^{\chi}(y) - h_b(y)) > 0, \qquad (4.5)$$

and solve for the non-trivial optimal timing strategy.

4.1.2 *Optimal Switching Approach*

Under the optimal switching approach, it is assumed that an infinite number of entry and exit actions take place. The sequential entry and exit times are modeled by the stopping times $\nu_1, \tau_1, \nu_2, \tau_2, \cdots \in \mathcal{T}$ such that

$$0 \leq \nu_1 \leq \tau_1 \leq \nu_2 \leq \tau_2 \leq \dots.$$

Entry and exit decisions are made, respectively, at times ν_i and τ_i, $i \in \mathbb{N}$. The optimal timing to enter or exit would depend on the initial position. Precisely, under the CIR model, if the initial position is zero, then the first task is to determine when to *start* and the corresponding optimal switching problem is

$$\tilde{J}^{\chi}(y) = \sup_{\Lambda_0} \mathbb{E}_y \left\{ \sum_{n=1}^{\infty} [e^{-r\tau_n} h_s(Y_{\tau_n}) - e^{-r\nu_n} h_b(Y_{\nu_n})] \right\}, \qquad (4.6)$$

with the set of admissible stopping times $\Lambda_0 = (\nu_1, \tau_1, \nu_2, \tau_2, \dots)$, and the reward functions h_b and h_s defined in (4.4). On the other hand, if we start with a long position, then it is necessary to solve

$$\tilde{V}^{\chi}(y) = \sup_{\Lambda_1} \mathbb{E}_y \left\{ e^{-r\tau_1} h_s(Y_{\tau_1}) + \sum_{n=2}^{\infty} [e^{-r\tau_n} h_s(Y_{\tau_n}) - e^{-r\nu_n} h_b(Y_{\nu_n})] \right\}, \qquad (4.7)$$

with $\Lambda_1 = (\tau_1, \nu_2, \tau_2, \nu_3, \dots)$ to determine when to *stop*.

In summary, the optimal starting-stopping and switching problems differ in the number of entry and exit decisions. Observe that any strategy for the starting-stopping problem (4.2)–(4.3) is also a candidate strategy for the switching problem (4.6)–(4.7). Therefore, it follows that $V^\chi(y) \leq \tilde{V}^\chi(y)$ and $J^\chi(y) \leq \tilde{J}^\chi(y)$. Our objective is to derive and compare the corresponding optimal timing strategies under these two approaches.

4.2 Summary of Analytical Results

We first summarize our analytical results and illustrate the optimal starting and stopping strategies. The method of solutions and their proofs will be discussed in Section 4.3.

We consider the optimal starting-stopping problem followed by the optimal switching problem. First, we denote the infinitesimal generator of Y as

$$\mathcal{L}^\chi = \frac{\sigma^2 y}{2}\frac{d^2}{dy^2} + \mu(\theta - y)\frac{d}{dy},$$

and consider the ordinary differential equation (ODE)

$$\mathcal{L}^\chi u(y) = r u(y), \quad \text{for } y \in \mathbb{R}_+. \tag{4.8}$$

To present the solutions of this ODE, we define the functions

$$F^\chi(y) := M(\frac{r}{\mu}, \frac{2\mu\theta}{\sigma^2}; \frac{2\mu y}{\sigma^2}), \quad \text{and} \quad G^\chi(y) := U(\frac{r}{\mu}, \frac{2\mu\theta}{\sigma^2}; \frac{2\mu y}{\sigma^2}), \tag{4.9}$$

where

$$M(a, b; z) = \sum_{n=0}^{\infty} \frac{a_n z^n}{b_n n!}, \quad a_0 = 1, \ a_n = a(a+1)(a+2)\cdots(a+n-1),$$

$$U(a, b; z) = \frac{\Gamma(1-b)}{\Gamma(a-b+1)}M(a, b; z) + \frac{\Gamma(b-1)}{\Gamma(a)}z^{1-b}M(a-b+1, 2-b; z)$$

are the confluent hypergeometric functions of first and second kind, also called the Kummer's function and Tricomi's function, respectively (see Chapter 13 of Abramowitz and Stegun (1965) and Chapter 9 of Lebedev (1972)). As is well known (see Göing-Jaeschke and Yor (2003)), F^χ and G^χ are strictly positive and, respectively, the strictly increasing and decreasing continuously differentiable solutions of the ODE (4.8). Also, we remark that the discounted processes $(e^{-rt}F^\chi(Y_t))_{t \geq 0}$ and $(e^{-rt}G^\chi(Y_t))_{t \geq 0}$ are martingales.

In addition, recall the reward functions defined in (4.4) and note that

$$(\mathcal{L}^{\chi} - r)h_b(y) \begin{cases} > 0 & \text{if } y < y_b, \\ < 0 & \text{if } y > y_b, \end{cases} \tag{4.10}$$

and

$$(\mathcal{L}^{\chi} - r)h_s(y) \begin{cases} > 0 & \text{if } y < y_s, \\ < 0 & \text{if } y > y_s, \end{cases} \tag{4.11}$$

where the critical constants y_b and y_s are defined by

$$y_b := \frac{\mu\theta - rc_b}{\mu + r} \quad \text{and} \quad y_s := \frac{\mu\theta + rc_s}{\mu + r}. \tag{4.12}$$

Note that y_b and y_s depend on the parameters μ, θ and r, as well as c_b and c_s respectively, but not σ.

4.2.1 *Optimal Starting-Stopping Problem*

We now present the results for the optimal starting-stopping problem (4.2)–(4.3). As it turns out, the value function V^{χ} is expressed in terms of F^{χ}, and J^{χ} in terms of V^{χ} and G^{χ}. The functions F^{χ} and G^{χ} also play a role in determining the optimal starting and stopping thresholds.

First, we give a bound on the value function V^{χ} in terms of $F^{\chi}(y)$.

Lemma 4.1. *There exists a positive constant K^{χ} such that, for all $y \geq 0$, $0 \leq V^{\chi}(y) \leq K^{\chi}F^{\chi}(y)$.*

Theorem 4.2. *The value function for the optimal stopping problem (4.2) is given by*

$$V^{\chi}(y) = \begin{cases} \frac{b^{\chi*} - c_s}{F^{\chi}(b^{\chi*})} F^{\chi}(y) & \text{if } y \in [0, b^{\chi*}), \\ y - c_s & \text{if } y \in [b^{\chi*}, +\infty). \end{cases}$$

Here, the optimal stopping level $b^{\chi*} \in (c_s \vee y_s, \infty)$ is found from the equation

$$F^{\chi}(b) = (b - c_s)F^{\chi\prime}(b). \tag{4.13}$$

Therefore, it is optimal to stop as soon as the process Y reaches $b^{\chi*}$ from below. The stopping level $b^{\chi*}$ must also be higher than the fixed cost c_s as well as the critical level y_s defined in (4.12).

Now we turn to the optimal starting problem. Define the reward function

$$\hat{h}^{\chi}(y) := V^{\chi}(y) - (y + c_b). \tag{4.14}$$

Since F^X, and thus V^X, are convex, so is $\hat{h}^X$, we also observe that the reward function $\hat{h}^X(y)$ is decreasing in y. To exclude the scenario where it is optimal never to start, the condition stated in (4.5), namely, $\sup_{y \in \mathbb{R}+} \hat{h}^X(y) > 0$, is now equivalent to

$$V^X(0) = \frac{b^{X*} - c_s}{F^X(b^{X*})} > c_b, \tag{4.15}$$

since $F^X(0) = 1$.

Lemma 4.3. *For all $y \geq 0$, the value function satisfies the inequality $0 \leq J^X(y) \leq (\frac{b^{X*}-c_s}{F^X(b^{X*})} - c_b)^+$.*

Theorem 4.4. *The optimal starting problem (4.3) admits the solution*

$$J^X(y) = \begin{cases} V^X(y) - (y + c_b) & \text{if } y \in [0, d^{X*}], \\ \frac{V^X(d^{X*}) - (d^{X*} + c_b)}{G^X(d^{X*})} G^X(y) & \text{if } y \in (d^{X*}, +\infty). \end{cases}$$

The optimal starting level $d^{X*} > 0$ is uniquely determined from

$$G^X(d)(V^{X'}(d) - 1) = G^{X'}(d)(V^X(d) - (d + c_b)). \tag{4.16}$$

As a result, it is optimal to start as soon as the CIR process Y falls below the strictly positive level d^{X*}.

4.2.2   *Optimal Switching Problem*

Now we study the optimal switching problem under the CIR model in (4.1).

Lemma 4.5. *For all $y \geq 0$, the value functions $\tilde{J}^X$ and $\tilde{V}^X$ satisfy the inequalities*

$$0 \leq \tilde{J}^X(y) \leq \frac{\mu\theta}{r},$$

$$0 \leq \tilde{V}^X(y) \leq y + \frac{2\mu\theta}{r}.$$

We start by giving conditions under which it is optimal not to start ever.

Theorem 4.6. *Under the CIR model, if it holds that*

(i) $y_b \leq 0$,   *or*
(ii) $y_b > 0$ *and* $c_b \geq \frac{b^{X*}-c_s}{F^X(b^{X*})}$,

with $b^{\chi*}$ given in (4.13), then the optimal switching problem (4.6)–(4.7) admits the solution

$$\tilde{J}^\chi(y) = 0 \qquad for \quad y \ge 0, \tag{4.17}$$

and

$$\tilde{V}^\chi(y) = \begin{cases} \frac{b^{\chi*}-c_s}{F^\chi(b^{\chi*})}F^\chi(y) & if\ y \in [0, b^{\chi*}), \\ y - c_s & if\ y \in [b^{\chi*}, +\infty). \end{cases} \tag{4.18}$$

Conditions (i) and (ii) depend on problem data and can be easily verified. In particular, recall that y_b is defined in (4.12) and is easy to compute, furthermore it is independent of σ and c_s. Since it is optimal to never enter, the switching problem is equivalent to a stopping problem and the solution in Theorem 4.6 agrees with that in Theorem 4.2. Next, we provide conditions under which it is optimal to enter as soon as the CIR process reaches some lower level.

Theorem 4.7. *Under the CIR model, if*

$$y_b > 0 \quad and \quad c_b < \frac{b^{\chi*} - c_s}{F^\chi(b^{\chi*})}, \tag{4.19}$$

with $b^{\chi*}$ given in (4.13), then the optimal switching problem (4.6)–(4.7) admits the solution

$$\tilde{J}^\chi(y) = \begin{cases} P^\chi F^\chi(y) - (y + c_b) & if\ y \in [0, \tilde{d}^{\chi*}], \\ Q^\chi G^\chi(y) & if\ y \in (\tilde{d}^{\chi*}, +\infty), \end{cases} \tag{4.20}$$

and

$$\tilde{V}^\chi(y) = \begin{cases} P^\chi F^\chi(y) & if\ y \in [0, \tilde{b}^{\chi*}), \\ Q^\chi G^\chi(y) + (y - c_s) & if\ y \in [\tilde{b}^{\chi*}, +\infty), \end{cases} \tag{4.21}$$

where

$$P^\chi = \frac{G^\chi(\tilde{d}^{\chi*}) - (\tilde{d}^{\chi*} + c_b)G^{\chi'}(\tilde{d}^{\chi*})}{F^{\chi'}(\tilde{d}^{\chi*})G^\chi(\tilde{d}^{\chi*}) - F^\chi(\tilde{d}^{\chi*})G^{\chi'}(\tilde{d}^{\chi*})},$$

$$Q^\chi = \frac{F^\chi(\tilde{d}^{\chi*}) - (\tilde{d}^{\chi*} + c_b)F^{\chi'}(\tilde{d}^{\chi*})}{F^{\chi'}(\tilde{d}^{\chi*})G^\chi(\tilde{d}^{\chi*}) - F^\chi(\tilde{d}^{\chi*})G^{\chi'}(\tilde{d}^{\chi*})}.$$

There exist unique optimal starting and stopping levels $\tilde{d}^{\chi*}$ and $\tilde{b}^{\chi*}$, which are found from the nonlinear system of equations:

$$\frac{G^\chi(d) - (d + c_b)G^{\chi'}(d)}{F^{\chi'}(d)G^\chi(d) - F^\chi(d)G^{\chi'}(d)} = \frac{G^\chi(b) - (b - c_s)G^{\chi'}(b)}{F^{\chi'}(b)G^\chi(b) - F^\chi(b)G^{\chi'}(b)},$$

$$\frac{F^\chi(d) - (d + c_b)F^{\chi'}(d)}{F^{\chi'}(d)G^\chi(d) - F^\chi(d)G^{\chi'}(d)} = \frac{F^\chi(b) - (b - c_s)F^{\chi'}(b)}{F^{\chi'}(b)G^\chi(b) - F^\chi(b)G^{\chi'}(b)}.$$

Moreover, we have that $\tilde{d}^{\chi*} < y_b$ and $\tilde{b}^{\chi*} > y_s$.

In this case, it is optimal to start and stop an infinite number of times where we start as soon as the CIR process drops to $\tilde{d}^{\chi*}$ and stop when the process reaches $\tilde{b}^{\chi*}$. Note that in the case of Theorem 4.6 where it is never optimal to start, the optimal stopping level $b^{\chi*}$ is the same as that of the optimal stopping problem in Theorem 4.2. The optimal starting level $\tilde{d}^{\chi*}$, which only arises when it is optimal to start and stop sequentially, is in general not the same as $d^{\chi*}$ in Theorem 4.4.

We conclude the section with two remarks.

Remark 4.8. Given the model parameters, in order to identify which of Theorem 4.6 or Theorem 4.7 applies, we begin by checking whether $y_b \leq 0$. If so, it is optimal not to enter. Otherwise, Theorem 4.6 still applies if $c_b \geq \frac{b^{\chi*} - c_s}{F^\chi(b^{\chi*})}$ holds. In the other remaining case, the problem is solved as in Theorem 4.7. In fact, the condition $c_b < \frac{b^{\chi*} - c_s}{F^\chi(b^{\chi*})}$ implies $y_b > 0$ (see the proof of Lemma 4.12 in the Appendix). Therefore, condition (4.19) in Theorem 4.7 is in fact identical to (4.15) in Theorem 4.4.

Remark 4.9. To verify the optimality of the results in Theorems 4.6 and 4.7, one can show by direct substitution that the solutions $(\tilde{J}^\chi, \tilde{V}^\chi)$ in (4.17)–(4.18) and (4.20)–(4.21) satisfy the variational inequalities:

$$\min\{r\tilde{J}^\chi(y) - \mathcal{L}^\chi \tilde{J}^\chi(y), \tilde{J}^\chi(y) - (\tilde{V}^\chi(y) - (y + c_b))\} = 0,$$
$$\min\{r\tilde{V}^\chi(y) - \mathcal{L}^\chi \tilde{V}^\chi(y), \tilde{V}^\chi(y) - (\tilde{J}^\chi(y) + (y - c_s))\} = 0.$$

Indeed, this is the approach used by Zervos *et al.* (2013) for checking the solutions of their optimal switching problems.

4.2.3 *Numerical Examples*

We numerically implement Theorems 4.2, 4.4, and 4.7, and illustrate the associated starting and stopping thresholds. In Figure 4.2(a), we observe the changes in optimal starting and stopping levels as speed of mean reversion increases. Both starting levels $d^{\chi*}$ and $\tilde{d}^{\chi*}$ rise with μ, from 0.0964 to 0.1219 and from 0.1460 to 0.1696, respectively, as μ increases from 0.3 to 0.85. The optimal switching stopping level $\tilde{b}^{\chi*}$ also increases. On the other hand, stopping level $b^{\chi*}$ for the starting-stopping problem stays relatively constant as μ changes.

Figure 4.1 shows a simulated CIR path along with optimal entry and exit levels for both starting-stopping and switching problems. Under the starting-stopping problem, it is optimal to start once the process reaches $d^{\chi*} = 0.0373$ and to stop when the process hits $b^{\chi*} = 0.4316$. For the switching problem, it is optimal to start once the process values hits $\tilde{d}^{\chi*} = 0.1189$ and to stop when the value of the CIR process rises to $\tilde{b}^{\chi*} = 0.2078$. We note that both stopping levels $b^{\chi*}$ and $\tilde{b}^{\chi*}$ are higher than the long-run mean $\theta = 0.2$, and the starting levels $d^{\chi*}$ and $\tilde{d}^{\chi*}$ are lower than θ. The process starts at $Y_0 = 0.15 > \tilde{d}^{\chi*}$, under the optimal switching setting, the first time to enter occurs on day 8 when the process falls to 0.1172 and subsequently exits on day 935 at a level of 0.2105. For the starting-stopping problem, entry takes place much later on day 200 when the process hits 0.0306 and exits on day 2671 at 0.4369. Under the optimal switching problem, two entries and two exits will be completed by the time a single entry-exit sequence is realized for the starting-stopping problem.

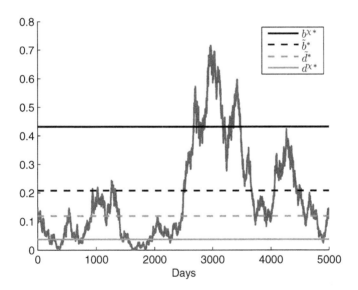

Fig. 4.1   A sample CIR path, along with starting and stopping levels. Under the starting-stopping setting, a starting decision is made at $\nu_{d^{\chi*}} = \inf\{t \geq 0 : Y_t \leq d^{\chi*} = 0.0373\}$, and a stopping decision is made at $\tau_{b^{\chi*}} = \inf\{t \geq \nu_{d^{\chi*}} : Y_t \geq b^{\chi*} = 0.4316\}$. Under the optimal switching problem, entry and exit take place at $\nu_{\tilde{d}^{\chi*}} = \inf\{t \geq 0 : Y_t \leq \tilde{d}^{\chi*} = 0.1189\}$, and $\tau_{\tilde{b}^{\chi*}} = \inf\{t \geq \nu_{\tilde{d}^{\chi*}} : Y_t \geq \tilde{b}^{\chi*} = 0.2078\}$ respectively. Parameters: $\mu = 0.2$, $\sigma = 0.3$, $\theta = 0.2$, $r = 0.05$, $c_s = 0.001$, $c_b = 0.001$.

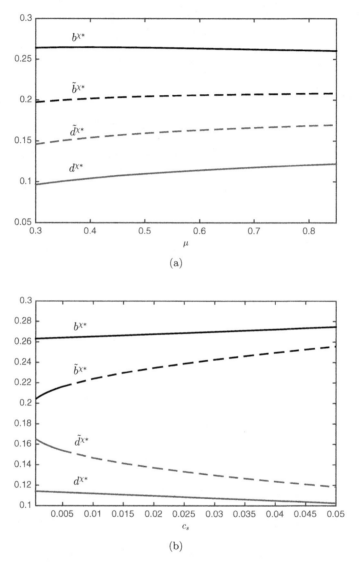

Fig. 4.2 (a) The optimal starting and stopping levels vs speed of mean reversion μ. Parameters: $\sigma = 0.15$, $\theta = 0.2$, $r = 0.05$, $c_s = 0.001$, $c_b = 0.001$. (b) The optimal starting and stopping levels vs transaction cost c_s. Parameters: $\mu = 0.6$, $\sigma = 0.15$, $\theta = 0.2$, $r = 0.05$, $c_b = 0.001$.

In Figure 4.2(b), we see that as the stopping cost c_s increases, the increase in the optimal stopping levels is accompanied by a fall in optimal starting levels. In particular, the stopping levels, $b^{\chi*}$ and $\tilde{b}^{\chi*}$ increase. In comparison, both starting levels $d^{\chi*}$ and $\tilde{d}^{\chi*}$ fall. The lower starting level and higher stopping level mean that the entry and exit times are both delayed as a result of a higher transaction cost. Interestingly, although the cost c_s applies only when the process is stopped, it also has an impact on the timing to *start*, as seen in the changes in $d^{\chi*}$ and $\tilde{d}^{\chi*}$ in the figure.

As shown in Figure 4.2, the continuation (waiting) region of the switching problem $(\tilde{d}^{\chi*}, \tilde{b}^{\chi*})$ lies within that of the starting-stopping problem $(d^{\chi*}, b^{\chi*})$. The ability to enter and exit multiple times means it is possible to earn a smaller reward on each individual start-stop sequence while maximizing aggregate return. Moreover, we observe that optimal entry and exit levels of the starting-stopping problem is less sensitive to changes in model parameters than the entry and exit thresholds of the switching problem.

4.3 Methods of Solution and Proofs

We now provide detailed proofs for our analytical results in Section 4.2 beginning with the optimal starting-stopping problem. Our main result here is Theorem 4.10 which provides a mathematical characterization of the value function, and establishes the optimality of our method of constructing the solution.

4.3.1 *Optimal Starting-Stopping Problem*

We first describe the general solution procedure for the stopping problem V^{χ}, followed by the starting problem J^{χ}.

4.3.1.1 *Optimal Stopping Timing*

A key step of our solution method involves the transformation

$$\phi(y) := -\frac{G^{\chi}(y)}{F^{\chi}(y)}, \quad y \geq 0. \tag{4.22}$$

With this, we also define the function

$$H^{\chi}(z) := \begin{cases} \frac{h_s}{F^{\chi}} \circ \phi^{-1}(z) & \text{if } z < 0, \\ \lim\limits_{y \to +\infty} \frac{(h_s(y))^+}{F^{\chi}(y)} & \text{if } z = 0, \end{cases} \tag{4.23}$$

where h_s is given in (4.4). We now prove the analytical form for the value function.

Theorem 4.10. *Under the CIR model, the value function V^χ of (4.2) is given by*

$$V^\chi(y) = F^\chi(y)W^\chi(\phi(y)), \qquad (4.24)$$

with F^χ and ϕ given in (4.9) and (4.22) respectively, and W^χ is the decreasing smallest concave majorant of H^χ in (4.23).

Proof. We first show that $V^\chi(y) \geq F^\chi(y)W^\chi(\phi(y))$. Start at any $y \in [0, +\infty)$, we consider the first stopping time of Y from an interval $[a, b]$ with $0 \leq a \leq y \leq b \leq +\infty$. We compute the corresponding expected discounted reward

$$\mathbb{E}_y\{e^{-r(\tau_a \wedge \tau_b)} h_s(Y_{\tau_a \wedge \tau_b})\}$$
$$= h_s(a)\mathbb{E}_y\{e^{-r\tau_a}\mathbf{1}_{\{\tau_a < \tau_b\}}\} + h_s(b)\mathbb{E}_y\{e^{-r\tau_b}\mathbf{1}_{\{\tau_a > \tau_b\}}\}$$
$$= h_s(a)\frac{F^\chi(y)G^\chi(b) - F^\chi(b)G^\chi(y)}{F^\chi(a)G^\chi(b) - F^\chi(b)G^\chi(a)} + h_s(b)\frac{F^\chi(a)G^\chi(y) - F^\chi(y)G^\chi(a)}{F^\chi(a)G^\chi(b) - F^\chi(b)G^\chi(a)}$$
$$= F^\chi(y)\left[\frac{h_s(a)}{F^\chi(a)}\frac{\phi(b) - \phi(y)}{\phi(b) - \phi(a)} + \frac{h_s(b)}{F^\chi(b)}\frac{\phi(y) - \phi(a)}{\phi(b) - \phi(a)}\right]$$
$$= F^\chi(\phi^{-1}(z))\left[H^\chi(z_a)\frac{z_b - z}{z_b - z_a} + H^\chi(z_b)\frac{z - z_a}{z_b - z_a}\right],$$

where $z_a = \phi(a)$, $z_b = \phi(b)$.

Since $V^\chi(y) \geq \sup_{\{a,b:a \leq y \leq b\}} \mathbb{E}_y\{e^{-r(\tau_a \wedge \tau_b)} h_s(Y_{\tau_a \wedge \tau_b})\}$, we have

$$\frac{V^\chi(\phi^{-1}(z))}{F^\chi(\phi^{-1}(z))} \geq \sup_{\{z_a,z_b:z_a \leq z \leq z_b\}}\left[H^\chi(z_a)\frac{z_b - z}{z_b - z_a} + H^\chi(z_b)\frac{z - z_a}{z_b - z_a}\right],$$

$$(4.25)$$

which implies that $V^\chi(\phi^{-1}(z))/F^\chi(\phi^{-1}(z))$ dominates the concave majorant of H^χ.

Under the CIR model, the class of interval-type strategies does not include all single threshold-type strategies. In particular, the minimum value that a can take is 0. If $2\mu\theta < \sigma^2$, then Y can reach level 0 and reflects. The interval-type strategy with $a = 0$ implies stopping the process Y at level 0, even though it could be optimal to wait and let Y evolve.

Hence, we must also consider separately the candidate strategy of waiting for Y to reach an upper level $b \geq y$ without a lower stopping level. The well-known supermartingale property of $(e^{-rt}V^\chi(Y_t))_{t\geq 0}$ (see Appendix D of Karatzas and Shreve (1998)) implies that $V^\chi(y) \geq \mathbb{E}_y\{e^{-r\tau}V^\chi(Y_\tau)\}$ for $\tau \in \mathcal{T}$. Then, taking $\tau = \tau_b$, we have

$$V^\chi(y) \geq \mathbb{E}_y\{e^{-r\tau_b}V^\chi(Y_{\tau_b})\} = V^\chi(b)\frac{F^\chi(y)}{F^\chi(b)},$$

or equivalently,

$$\frac{V^\chi(\phi^{-1}(z))}{F^\chi(\phi^{-1}(z))} = \frac{V^\chi(y)}{F^\chi(y)} \geq \frac{V^\chi(b)}{F^\chi(b)} = \frac{V^\chi(\phi^{-1}(z_b))}{F^\chi(\phi^{-1}(z_b))}, \tag{4.26}$$

which indicates that $V^\chi(\phi^{-1}(z))/F^\chi(\phi^{-1}(z))$ is *decreasing*. By (4.25) and (4.26), we now see that $V^\chi(y) \geq F^\chi(y)W^\chi(\phi(y))$, where W^χ is the *decreasing* smallest concave majorant of H^χ.

For the reverse inequality, we first show that

$$F^\chi(y)W^\chi(\phi(y)) \geq \mathbb{E}_y\{e^{-r(t\wedge\tau)}F^\chi(Y_{t\wedge\tau})W^\chi(\phi(Y_{t\wedge\tau}))\}, \tag{4.27}$$

for $y \in [0, +\infty)$, $\tau \in \mathcal{T}$ and $t \geq 0$. If the initial value $y = 0$, then the decreasing property of W^χ implies the inequality

$$\mathbb{E}_0\{e^{-r(t\wedge\tau)}F^\chi(Y_{t\wedge\tau})W^\chi(\phi(Y_{t\wedge\tau}))\} \leq \mathbb{E}_0\{e^{-r(t\wedge\tau)}F^\chi(Y_{t\wedge\tau})\}W^\chi(\phi(0))$$
$$= F^\chi(0)W^\chi(\phi(0)),$$

where the equality follows from the martingale property of $(e^{-rt}F^\chi(Y_t))_{t\geq0}$.

When $y > 0$, the concavity of W^χ implies that, for any fixed z, there exists an affine function $L_z^\chi(\alpha) := m_z^\chi\alpha + c_z^\chi$ such that $L_z^\chi(\alpha) \geq W^\chi(\alpha)$ for $\alpha \geq \phi(0)$ and $L_z^\chi(z) = W^\chi(z)$ at $\alpha = z$, with constants m_z^χ and c_z^χ. In turn, this yields the inequality

$$\mathbb{E}_y\{e^{-r(\tau_0\wedge t\wedge\tau)}F^\chi(Y_{\tau_0\wedge t\wedge\tau})W^\chi(\phi(Y_{\tau_0\wedge t\wedge\tau}))\} \tag{4.28}$$
$$\leq \mathbb{E}_y\{e^{-r(\tau_0\wedge t\wedge\tau)}F^\chi(Y_{\tau_0\wedge t\wedge\tau})L_{\phi(y)}^\chi(\phi(Y_{\tau_0\wedge t\wedge\tau}))\}$$
$$= m_{\phi(y)}^\chi\mathbb{E}_y\{e^{-r(\tau_0\wedge t\wedge\tau)}F^\chi(Y_{\tau_0\wedge t\wedge\tau})\phi(Y_{\tau_0\wedge t\wedge\tau})\}$$
$$\quad + c_{\phi(y)}^\chi\mathbb{E}_y\{e^{-r(\tau_0\wedge t\wedge\tau)}F^\chi(Y_{\tau_0\wedge t\wedge\tau})\}$$
$$= -m_{\phi(y)}^\chi\mathbb{E}_y\{e^{-r(\tau_0\wedge t\wedge\tau)}G^\chi(Y_{\tau_0\wedge t\wedge\tau})\}$$
$$\quad + c_{\phi(y)}^\chi\mathbb{E}_y\{e^{-r(\tau_0\wedge t\wedge\tau)}F^\chi(Y_{\tau_0\wedge t\wedge\tau})\}$$
$$= -m_{\phi(y)}^\chi G^\chi(y) + c_{\phi(y)}^\chi F^\chi(y) \tag{4.29}$$
$$= F^\chi(y)L_{\phi(y)}^\chi(\phi(y))$$
$$= F^\chi(y)W^\chi(\phi(y)), \tag{4.30}$$

where (4.29) follows from the martingale property of $(e^{-rt}F^\chi(Y_t))_{t\geq0}$ and $(e^{-rt}G^\chi(Y_t))_{t\geq0}$. If $2\mu\theta \geq \sigma^2$, then $\tau_0 = +\infty$ for $y > 0$. This immediately reduces (4.28)–(4.30) to the desired inequality (4.27).

On the other hand, if $2\mu\theta < \sigma^2$, then we decompose (4.28) into two terms:

$$\mathbb{E}_y\{e^{-r(\tau_0 \wedge t \wedge \tau)}F^\chi(Y_{\tau_0 \wedge t \wedge \tau})W^\chi(\phi(Y_{\tau_0 \wedge t \wedge \tau}))\}$$

$$= \underbrace{\mathbb{E}_y\{e^{-r(t \wedge \tau)}F^\chi(Y_{t \wedge \tau})W^\chi(\phi(Y_{t \wedge \tau}))\mathbf{1}_{\{t \wedge \tau \leq \tau_0\}}\}}_{\text{(I)}}$$

$$+ \underbrace{\mathbb{E}_y\{e^{-r\tau_0}F^\chi(Y_{\tau_0})W^\chi(\phi(Y_{\tau_0}))\mathbf{1}_{\{t \wedge \tau > \tau_0\}}\}}_{\text{(II)}}.$$

By the optional sampling theorem and decreasing property of W^χ, the second term satisfies

$$\text{(II)} = W^\chi(\phi(0))\mathbb{E}_y\{e^{-r\tau_0}F^\chi(Y_{\tau_0})\mathbf{1}_{\{t \wedge \tau > \tau_0\}}\}$$

$$\geq W^\chi(\phi(0))\mathbb{E}_y\{e^{-r(t \wedge \tau)}F^\chi(Y_{t \wedge \tau})\mathbf{1}_{\{t \wedge \tau > \tau_0\}}\}$$

$$\geq \mathbb{E}_y\{e^{-r(t \wedge \tau)}F^\chi(Y_{t \wedge \tau})W^\chi(\phi(Y_{t \wedge \tau}))\mathbf{1}_{\{t \wedge \tau > \tau_0\}}\} =: \text{(II')}. \quad (4.31)$$

Combining (4.31) with (4.30), we arrive at

$$F^\chi(y)W^\chi(\phi(y)) \geq \text{(I)} + \text{(II')} = \mathbb{E}_y\{e^{-r(t \wedge \tau)}F^\chi(Y_{t \wedge \tau})W^\chi(\phi(Y_{t \wedge \tau}))\},$$

for all $y > 0$. In all, inequality (4.27) holds for all $y \in [0, +\infty)$, $\tau \in \mathcal{T}$ and $t \geq 0$. From (4.27) and the fact that W^χ majorizes H^χ, it follows that

$$F^\chi(y)W^\chi(\phi(y)) \geq \mathbb{E}_y\{e^{-r(t \wedge \tau)}F^\chi(Y_{t \wedge \tau})W^\chi(\phi(Y_{t \wedge \tau}))\}$$

$$\geq \mathbb{E}_y\{e^{-r(t \wedge \tau)}F^\chi(Y_{t \wedge \tau})H^\chi(\phi(Y_{t \wedge \tau}))\}$$

$$\geq \mathbb{E}_y\{e^{-r(t \wedge \tau)}h_s(Y_{t \wedge \tau})\}. \quad (4.32)$$

Maximizing (4.32) over all $\tau \in \mathcal{T}$ and $t \geq 0$ yields the reverse inequality $F^\chi(y)W^\chi(\phi(y)) \geq V^\chi(y)$. $\qquad\square$

In summary, we have found an expression for the value function $V^\chi(y)$ in (4.24), and proved that it is sufficient to consider only candidate stopping times described by the first time Y reaches a single upper threshold or exits an interval. To determine the optimal timing strategy, we need to understand the properties of H^χ and its smallest concave majorant W^χ. To this end, we have the following lemma.

Lemma 4.11. *The function H^χ is continuous on $[\phi(0), 0]$, twice differentiable on $(\phi(0), 0)$ and possesses the following properties:*

(i) $H^\chi(0) = 0$, and

$$H^\chi(z) \begin{cases} < 0 & \text{if } z \in [\phi(0), \phi(c_s)), \\ > 0 & \text{if } z \in (\phi(c_s), 0). \end{cases} \quad (4.33)$$

(ii) $H^\chi(z)$ *is strictly increasing for* $z \in (\phi(0), \phi(c_s) \vee \phi(y_s))$.

(iii)

$$H^\chi(z) \ is \ \begin{cases} convex & if \ z \in (\phi(0), \phi(y_s)], \\ concave & if \ z \in [\phi(y_s), 0). \end{cases}$$

In Figure 4.3, we see that H^χ is first increasing then decreasing, and first convex then concave. Using these properties, we now derive the optimal stopping timing.

Proof of Theorem 4.2 We determine the value function in the form: $V^\chi(y) = F^\chi(y)W^\chi(\phi(y))$, where W^χ is the decreasing smallest concave majorant of H^χ. By Lemma 4.11 and Figure 4.3, H^χ peaks at $z^{\chi*} > \phi(c_s) \vee \phi(y_s)$ so that

$$H^{\chi\prime}(z^{\chi*}) = 0. \tag{4.34}$$

In turn, the decreasing smallest concave majorant admits the form:

$$W^\chi(z) = \begin{cases} H^\chi(z^{\chi*}) & if \ z < z^{\chi*}, \\ H^\chi(z) & if \ z \geq z^{\chi*}. \end{cases} \tag{4.35}$$

Substituting $b^{\chi*} = \phi^{-1}(z^{\chi*})$ into (4.34), we have

$$\begin{aligned} H^{\chi\prime}(z^{\chi*}) &= \frac{F^\chi(\phi^{-1}(z^{\chi*})) - (\phi^{-1}(z^{\chi*}) - c_s)F^{\chi\prime}(\phi^{-1}(z^{\chi*}))}{F^{\chi\prime}(\phi^{-1}(z^{\chi*}))G^\chi(\phi^{-1}(z^{\chi*})) - F^\chi(\phi^{-1}(z^{\chi*}))G^{\chi\prime}(\phi^{-1}(z^{\chi*}))} \\ &= \frac{F^\chi(b^{\chi*}) - (b^{\chi*} - c_s)F^{\chi\prime}(b^{\chi*})}{F^{\chi\prime}(b^{\chi*})G^\chi(b^{\chi*}) - F^\chi(b^{\chi*})G^{\chi\prime}(b^{\chi*})}, \end{aligned}$$

which can be further simplified to (4.13). We can express $H^\chi(z^{\chi*})$ in terms of $b^{\chi*}$:

$$H^\chi(z^{\chi*}) = \frac{b^{\chi*} - c_s}{F^\chi(b^{\chi*})}. \tag{4.36}$$

Applying (4.36) to (4.35), we get

$$W^\chi(\phi(y)) = \begin{cases} H^\chi(z^{\chi*}) = \frac{b^{\chi*} - c_s}{F^\chi(b^{\chi*})} & if \ y < b^{\chi*}, \\ H^\chi(\phi(y)) = \frac{y - c_s}{F^\chi(y)} & if \ y \geq b^{\chi*}. \end{cases}$$

Finally, substituting this into the value function $V^\chi(y) = F^\chi(y)W^\chi(\phi(y))$, we conclude.

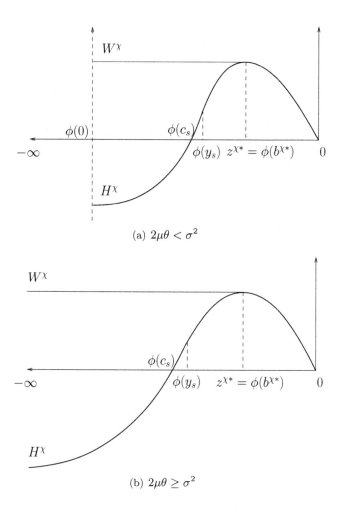

Fig. 4.3   Sketches of H^χ and W^χ. The function W^χ is equal to the constant $H^\chi(z^{\chi*})$ on $(\phi(0), z^{\chi*})$, and coincides with H^χ on $[z^{\chi*}, 0]$. Note that $-\infty < \phi(0) < 0$ if $2\mu\theta < \sigma^2$, and $\phi(0) = -\infty$ if $2\mu\theta \geq \sigma^2$.

4.3.1.2   *Optimal Starting Timing*

We now turn to the optimal starting problem. Our methodology in Section 4.3.1.1 applies to general payoff functions, and thus can be applied to the optimal starting problem (4.3) as well. To this end, we apply the same transformation (4.22) and define the function

$$\hat{H}^{\chi}(z) := \begin{cases} \frac{\hat{h}^{\chi}}{F^{\chi}} \circ \phi^{-1}(z) & \text{if } z < 0, \\ \lim\limits_{y \to +\infty} \frac{(\hat{h}^{\chi}(y))^{+}}{F^{\chi}(y)} & \text{if } z = 0, \end{cases}$$

where $\hat{h}^{\chi}$ is given in (4.14). We then follow Theorem 4.2 to determine the value function J^{χ}. This amounts to finding the *decreasing* smallest concave majorant $\hat{W}^{\chi}$ of $\hat{H}^{\chi}$. Indeed, we can replace H^{χ} and W^{χ} with $\hat{H}^{\chi}$ and $\hat{W}^{\chi}$ in Theorem 4.2 and its proof. As a result, the value function of the optimal starting timing problem must take the form

$$J^{\chi}(y) = F^{\chi}(y)\hat{W}^{\chi}(\phi(y)).$$

To solve the optimal starting timing problem, we need to understand the properties of $\hat{H}^{\chi}$.

Lemma 4.12. *The function $\hat{H}^{\chi}$ is continuous on $[\phi(0), 0]$, differentiable on $(\phi(0), 0)$, and twice differentiable on $(\phi(0), \phi(b^{\chi*})) \cup (\phi(b^{\chi*}), 0)$, and possesses the following properties:*

(i) $\hat{H}^{\chi}(0) = 0$. Let $\bar{d}^{\chi}$ denote the unique solution to $\hat{h}^{\chi}(y) = 0$, then $\bar{d}^{\chi} < b^{\chi*}$ and

$$\hat{H}^{\chi}(z) \begin{cases} > 0 & \text{if } z \in [\phi(0), \phi(\bar{d}^{\chi})), \\ < 0 & \text{if } z \in (\phi(\bar{d}^{\chi}), 0). \end{cases}$$

(ii) $\hat{H}^{\chi}(z)$ is strictly increasing for $z > \phi(b^{\chi*})$ and $\lim_{z \to \phi(0)} \hat{H}^{\chi'}(z) = 0$.
(iii)

$$\hat{H}^{\chi}(z) \text{ is } \begin{cases} concave & \text{if } z \in (\phi(0), \phi(y_b)), \\ convex & \text{if } z \in (\phi(y_b), 0). \end{cases}$$

By Lemma 4.12, we sketch $\hat{H}^{\chi}$ in Figure 4.4.

Proof of Theorem 4.4 To determine the value function in the form: $J^{\chi}(y) = F^{\chi}(y)\hat{W}^{\chi}(\phi(y))$, we analyze the decreasing smallest concave majorant, $\hat{W}^{\chi}$, of $\hat{H}^{\chi}$. By Lemma 4.12 and Figure 4.3, we have $\hat{H}^{\chi'}(z) \to 0$ as $z \to \phi(0)$. Therefore, there exists a unique number $\hat{z}^{\chi} \in (\phi(0), \phi(b^{\chi*}))$ such that

$$\frac{\hat{H}^{\chi}(\hat{z}^{\chi})}{\hat{z}^{\chi}} = \hat{H}^{\chi'}(\hat{z}^{\chi}). \tag{4.37}$$

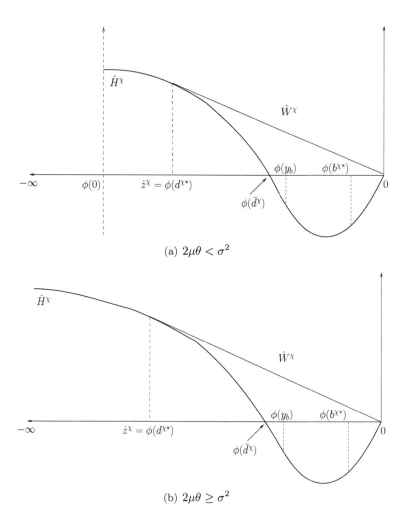

Fig. 4.4   Sketches of $\hat{H}^\chi$ and $\hat{W}^\chi$. The function $\hat{W}^\chi$ coincides with $\hat{H}^\chi$ on $[\phi(0), \hat{z}^\chi]$ and is a straight line tangent to $\hat{H}^\chi$ at $\hat{z}^\chi$ on $(\hat{z}^\chi, 0]$. Note that $-\infty < \phi(0) < 0$ if $2\mu\theta < \sigma^2$, and $\phi(0) = -\infty$ if $2\mu\theta \geq \sigma^2$.

In turn, the decreasing smallest concave majorant admits the form:

$$\hat{W}^\chi(z) = \begin{cases} \hat{H}^\chi(z) & \text{if } z \leq \hat{z}^\chi, \\ z\dfrac{\hat{H}^\chi(\hat{z}^\chi)}{\hat{z}^\chi} & \text{if } z > \hat{z}^\chi. \end{cases} \qquad (4.38)$$

Substituting $d^{\chi*} = \phi^{-1}(\hat{z}^\chi)$ into (4.37), we have

$$\frac{\hat{H}^\chi(\hat{z}^\chi)}{\hat{z}^\chi} = \frac{\hat{H}^\chi(\phi(d^{\chi*}))}{\phi(d^{\chi*})} = -\frac{V^\chi(d^{\chi*}) - d^{\chi*} - c_b}{G^\chi(d^{\chi*})}, \tag{4.39}$$

and

$$\hat{H}^{\chi'}(\hat{z}^\chi) = \frac{F^\chi(d^{\chi*})(V^{\chi'}(d^{\chi*}) - 1) - F^{\chi'}(d^{\chi*})(V^\chi(d^{\chi*}) - (d^{\chi*} + c_b))}{F^{\chi'}(d^{\chi*})G^\chi(d^{\chi*}) - F^\chi(d^{\chi*})G^{\chi'}(d^{\chi*})}.$$

Equivalently, we can express condition (4.37) in terms of $d^{\chi*}$:

$$-\frac{V^\chi(d^{\chi*}) - (d^{\chi*} + c_b)}{G^\chi(d^{\chi*})}$$

$$= \frac{F^\chi(d^{\chi*})(V^{\chi'}(d^{\chi*}) - 1) - F^{\chi'}(d^{\chi*})(V^\chi(d^{\chi*}) - (d^{\chi*} + c_b))}{F^{\chi'}(d^{\chi*})G^\chi(d^{\chi*}) - F^\chi(d^{\chi*})G^{\chi'}(d^{\chi*})},$$

which shows $d^{\chi*}$ satisfies (4.16) after simplification.

Applying (4.39) to (4.38), we get

$$W^\chi(\phi(y)) = \begin{cases} \hat{H}^\chi(\phi(y)) = \frac{V^\chi(y) - (y + c_b)}{F^\chi(y)} & \text{if } y \in [0, d^{\chi*}], \\ \phi(y)\frac{\hat{H}^\chi(\hat{z}^\chi)}{\hat{z}^\chi} = \frac{V^\chi(d^{\chi*}) - (d^{\chi*} + c_b)}{G^\chi(d^{\chi*})}\frac{G^\chi(y)}{F^\chi(y)} & \text{if } y \in (d^{\chi*}, +\infty). \end{cases}$$

From this, we obtain the value function.

4.3.2 Optimal Switching Problem

Proofs of Theorems 4.6 and 4.7 Zervos *et al.* (2013) have studied a similar problem of trading a mean-reverting asset with fixed transaction costs, and provided detailed proofs using a variational inequalities approach. In particular, we observe that y_b and y_s in (4.10) and (4.11) play the same roles as x_b and x_s in Assumption 4 in Zervos *et al.* (2013), respectively. However, Assumption 4 in Zervos *et al.* (2013) requires that $0 \le x_b$, this is not necessarily true for y_b in our problem. We have checked and realized that this assumption is not necessary for Theorem 4.6, and that $y_b < 0$ simply implies that there is no optimal starting level, i.e. it is never optimal to start.

In addition, Zervos *et al.* (2013) assume (in their Assumption 1) that the hitting time of level 0 is infinite with probability 1. In comparison, we consider not only the CIR case where 0 is inaccessible, but also when the CIR process has a reflecting boundary at 0. In fact, we find that the proofs in Zervos *et al.* (2013) apply to both cases under the CIR model. Therefore, apart from relaxation of the aforementioned assumptions, the proofs of our Theorems 4.6 and 4.7 are the same as that of Lemmas 1 and 2 in Zervos *et al.* (2013) respectively.

4.4   Proofs of Lemmas

In this last section, we present the proofs for the properties of V^χ, J^χ, H^χ and $\hat{H}^\chi$.

Proof of Lemma 4.1 (Bounds of V^χ)

First, the limit

$$\limsup_{y \to +\infty} \frac{(h_s(y))^+}{F^\chi(y)} = \limsup_{y \to +\infty} \frac{y - c_s}{F^\chi(y)} = \limsup_{y \to +\infty} \frac{1}{F^{\chi\prime}(y)} = 0.$$

Therefore, there exists some y_0 such that $(h_s(y))^+ < F^\chi(y)$ for $y \in (y_0, +\infty)$. As for $y \le y_0$, $(h_s(y))^+$ is bounded above by the constant $(y_0 - c_s)^+$. As a result, we can always find a constant K^χ such that $(h_s(y))^+ \le K^\chi F^\chi(y)$ for all $y \in \mathbb{R}$.

By definition, the process $(e^{-rt} F^\chi(Y_t))_{t \ge 0}$ is a martingale. This implies, for every $y \in \mathbb{R}_+$ and $\tau \in \mathcal{T}$,

$$K^\chi F^\chi(y) = \mathbb{E}_y\{e^{-r\tau} K^\chi F^\chi(Y_\tau)\} \ge \mathbb{E}_y\{e^{-r\tau}(h_s(Y_\tau))^+\} \ge \mathbb{E}_y\{e^{-r\tau}h_s(Y_\tau)\}.$$

Therefore, $V^\chi(y) \le K^\chi F^\chi(y)$. Lastly, the choice of $\tau = +\infty$ as a candidate stopping time implies that $V^\chi(y) \ge 0$.

Proof of Lemma 4.3 (Bounds of J^χ)

As we pointed out in Section 4.3.1.2 that $\hat{h}^\chi(y)$ is decreasing in y, thus so is $(\hat{h}^\chi(y))^+$. We can conclude that $(\hat{h}^\chi(y))^+ \le (V^\chi(0) - c_b)^+ = (\frac{b^{\chi*} - c_s}{F^\chi(b^{\chi*})} - c_b)^+$. The rest of the proof is similar to that of Lemma 3.3, with $\hat{K}$ changed to $(\frac{b^{\chi*} - c_s}{F^\chi(b^{\chi*})} - c_b)^+$.

Proof of Lemma 4.5 (Bounds of $\tilde{J}^\chi$ and $\tilde{V}^\chi$)

By definition, both $\tilde{J}^\chi(y)$ and $\tilde{V}^\chi(y)$ are nonnegative. Using Dynkin's formula, we have

$$\mathbb{E}_y\left\{e^{-r\tau_n} Y_{\tau_n}\right\} - \mathbb{E}_y\left\{e^{-r\nu_n} Y_{\nu_n}\right\} = \mathbb{E}_y\left\{\int_{\nu_n}^{\tau_n} e^{-rt}(\mathcal{L}^\chi - r)Y_t dt\right\}$$

$$= \mathbb{E}_y\left\{\int_{\nu_n}^{\tau_n} e^{-rt}\left(\mu\theta - (r + \mu)Y_t\right) dt\right\}.$$

For $y \ge 0$, the function $\mu\theta - (r + \mu)y$ is bounded by $\mu\theta$. It follows that

$$\mathbb{E}_y\left\{e^{-r\tau_n} Y_{\tau_n}\right\} - \mathbb{E}_y\left\{e^{-r\nu_n} Y_{\nu_n}\right\} \le \mu\theta\mathbb{E}_y\left\{\int_{\nu_n}^{\tau_n} e^{-rt} dt\right\}.$$

Since $y - c_s \leq y$ and $y + c_b \geq y$, we have

$$\mathbb{E}_y \left\{ \sum_{n=1}^{\infty} [e^{-r\tau_n} h_s(Y_{\tau_n}) - e^{-r\nu_n} h_b(Y_{\nu_n})] \right\}$$

$$\leq \sum_{n=1}^{\infty} \left(\mathbb{E}_y \left\{ e^{-r\tau_n} Y_{\tau_n} \right\} - \mathbb{E}_y \left\{ e^{-r\nu_n} Y_{\nu_n} \right\} \right)$$

$$\leq \sum_{n=1}^{\infty} \mu\theta \mathbb{E}_y \left\{ \int_{\nu_n}^{\tau_n} e^{-rt} dt \right\} \leq \mu\theta \int_0^{\infty} e^{-rt} dt = \frac{\mu\theta}{r}.$$

This implies that $0 \leq \tilde{J}^\chi(y) \leq \frac{\mu\theta}{r}$. Similarly,

$$\mathbb{E}_y \left\{ e^{-r\tau_1} h_s(Y_{\tau_1}) + \sum_{n=2}^{\infty} [e^{-r\tau_n} h_s(Y_{\tau_n}) - e^{-r\tau_n} h_b(Y_{\tau_n})] \right\}$$

$$\leq \frac{\mu\theta}{r} + \mathbb{E}_y \left\{ e^{-r\tau_1} h_b(Y_{\tau_1}) \right\}.$$

Letting $\nu_1 = 0$ and using Dynkin's formula again, we have

$$\mathbb{E}_y \left\{ e^{-r\tau_1} Y_{\tau_1} \right\} - y \leq \frac{\mu\theta}{r}.$$

This implies that

$$\tilde{V}^\chi(y) \leq \frac{\mu\theta}{r} + y + \frac{\mu\theta}{r} := y + \frac{2\mu\theta}{r}.$$

Proof of Lemma 4.11 (Properties of H^χ)

(i) First, we compute

$$H^\chi(0) = \lim_{y \to +\infty} \frac{(h_s(y))^+}{F^\chi(y)} = \lim_{y \to +\infty} \frac{y - c_s}{F^\chi(x)} = \lim_{y \to +\infty} \frac{1}{F^{\chi\prime}(y)} = 0.$$

Using the facts that $F^\chi(y) > 0$ and $\phi(y)$ is a strictly increasing function, (4.33) follows.

(ii) We look at the first derivative of H^χ:

$$H^{\chi\prime}(z) = \frac{1}{\phi'(y)} \left(\frac{h_s}{F^\chi} \right)'(y) = \frac{1}{\phi'(y)} \frac{F^\chi(y) - (y - c_s)F^{\chi\prime}(y)}{F^{\chi2}(y)}, \quad z = \phi(y).$$

Since both $\phi'(y)$ and $F^{\chi2}(y)$ are positive, it remains to determine the sign of $F^\chi(y) - (y - c_s)F^{\chi\prime}(y)$. Since $F^{\chi\prime}(y) > 0$, we can equivalently check the sign of $v(y) := \frac{F^\chi(y)}{F^{\chi\prime}(y)} - (y - c_s)$. Note that $v'(y) = -\frac{F^\chi(y)F^{\chi\prime\prime}(y)}{(F^{\chi\prime}(y))^2} < 0$. Therefore, $v(y)$ is a strictly decreasing function. Also, it is clear that $v(c_s) > 0$ and $v(y_s) > 0$. Consequently, $v(y) > 0$ if $y < (c_s \vee y_s)$ and hence, $H^\chi(z)$ is strictly increasing if $z \in (\phi(0), \phi(c_s) \vee \phi(y_s))$.

(iii) By differentiation, we have

$$H^{\chi''}(z) = \frac{2}{\sigma^2 F^\chi(y)(\phi'(y))^2}(\mathcal{L}^\chi - r)h_s(y), \quad z = \phi(y).$$

Since $\sigma^2, F^\chi(y)$ and $(\phi'(y))^2$ are all positive, the convexity/concavity of H^χ depends on the sign of

$$(\mathcal{L}^\chi - r)h_s(y) = \mu(\theta - y) - r(y - c_s)$$

$$= (\mu\theta + rc_s) - (\mu + r)y \quad \begin{cases} \geq 0 & \text{if } y \in [0, y_s], \\ \leq 0 & \text{if } y \in [y_s, +\infty), \end{cases}$$

which implies property (iii).

Proof of Lemma 4.12 (Properties of $\hat{H}^\chi$)

It is straightforward to check that $V^\chi(y)$ is continuous and differentiable everywhere, and twice differentiable everywhere except at $y = b^{\chi*}$, and all these holds for $\hat{h}^\chi(y) = V^\chi(y) - (y + c_b)$. Both F^χ and ϕ are twice differentiable. In turn, the continuity and differentiability of $\hat{H}^\chi$ on $(\phi(0), 0)$ and twice differentiability of $\hat{H}^\chi$ on $(\phi(0), \phi(b^{\chi*})) \cup (\phi(b^{\chi*}), 0)$ follow.

To show the continuity of $\hat{H}^\chi$ at 0, we note that

$$\hat{H}^\chi(0) = \lim_{y \to +\infty} \frac{(\hat{h}^\chi(y))^+}{F^\chi(y)} = \lim_{y \to +\infty} \frac{0}{F^\chi(y)} = 0, \quad \text{and}$$

$$\lim_{z \to 0} \hat{H}^\chi(z) = \lim_{y \to +\infty} \frac{\hat{h}^\chi}{F^\chi}(y) = \lim_{y \to +\infty} \frac{-(c_s + c_b)}{F^\chi(y)} = 0.$$

From this, we conclude that $\hat{H}^\chi$ is also continuous at 0.

(i) First, for $y \in [b^{\chi*}, +\infty)$, $\hat{h}^\chi(y) \equiv -(c_s + c_b)$. For $y \in (0, b^{\chi*})$, we compute

$$V^{\chi'}(y) = \frac{b^{\chi*} - c_s}{F^\chi(b^{\chi*})}F^{\chi'}(y) = \frac{F^{\chi'}(y)}{F^{\chi'}(b^{\chi*})}, \quad \text{by (4.13)}.$$

Recall that $F^{\chi'}(y)$ is a strictly increasing function and $\hat{h}^\chi(y) = V^\chi(y) - y - c_b$. Differentiation yields

$$\hat{h}^{\chi'}(y) = V^{\chi'}(y) - 1 = \frac{F^{\chi'}(y)}{F^{\chi'}(b^{\chi*})} - 1 < \frac{F^{\chi'}(b^{\chi*})}{F^{\chi'}(b^{\chi*})} - 1 = 0, \quad y \in (0, b^{\chi*}),$$

which implies that $\hat{h}^\chi(y)$ is strictly decreasing for $y \in (0, b^{\chi*})$. On the other hand, $\hat{h}^\chi(0) > 0$ as we are considering the non-trivial case. Therefore, there exists a unique solution $\bar{d}^\chi < b^{\chi*}$ to $\hat{h}^\chi(y) = 0$, such that $\hat{h}^\chi(y) > 0$ for $y \in [0, \bar{d}^\chi)$, and $\hat{h}^\chi(y) < 0$ for $y \in (\bar{d}^\chi, +\infty)$.

With $\hat{H}^{\chi}(z) = (\hat{h}^{\chi}/F^{\chi}) \circ \phi^{-1}(z)$, the above properties of $\hat{h}^{\chi}$, along with the facts that $\phi(y)$ is strictly increasing and $F^{\chi}(y) > 0$, imply property (i).
(ii) With $z = \phi(y)$, for $y > b^{\chi*}$, $\hat{H}^{\chi}(z)$ is strictly increasing since

$$\hat{H}^{\chi'}(z) = \frac{1}{\phi'(y)} \left(\frac{\hat{h}^{\chi}}{F^{\chi}}\right)'(y) = \frac{1}{\phi'(y)} \left(\frac{-(c_s + c_b)}{F^{\chi}(y)}\right)' = \frac{1}{\phi'(y)} \frac{(c_s + c_b)F^{\chi'}(y)}{F^{\chi^2}(y)} > 0.$$

When $y \to 0$, because $(\frac{\hat{h}^{\chi}(y)}{F^{\chi}(y)})'$ is finite, but $\phi'(y) \to +\infty$, we have $\lim_{z \to \phi(0)} \hat{H}^{\chi'}(z) = 0$.
(iii) Consider the second derivative:

$$\hat{H}^{\chi''}(z) = \frac{2}{\sigma^2 F(y)(\phi'(y))^2} (\mathcal{L}^{\chi} - r)\hat{h}^{\chi}(y).$$

The positiveness of σ^2, $F^{\chi}(y)$ and $(\phi'(y))^2$ suggests that we inspect the sign of $(\mathcal{L}^{\chi} - r)\hat{h}^{\chi}(y)$:

$$(\mathcal{L}^{\chi} - r)\hat{h}^{\chi}(y)$$
$$= \frac{1}{2}\sigma^2 y V^{\chi''}(y) + \mu(\theta - y)V^{\chi'}(y) - \mu(\theta - y) - r(V^{\chi}(y) - (y + c_b))$$
$$= \begin{cases} (\mu + r)y - \mu\theta + rc_b & \text{if } y < b^{\chi*}, \\ r(c_s + c_b) > 0 & \text{if } y > b^{\chi*}. \end{cases}$$

Since $\mu, r > 0$ by assumption, $(\mathcal{L}^{\chi} - r)\hat{h}^{\chi}(y)$ is strictly increasing on $(0, b^{\chi*})$. Next, we show that $0 < y_b < y_s < b^{\chi*}$. By the fact that $F^{\chi'}(0) = \frac{r}{\mu\theta}$ and the assumption that $V^{\chi}(0) = \frac{b^{\chi*} - c_s}{F^{\chi}(b^{\chi*})} > c_b$, we have

$$V^{\chi'}(0) = \frac{b^{\chi*} - c_s}{F^{\chi}(b^{\chi*})} F^{\chi'}(0) = \frac{b^{\chi*} - c_s}{F^{\chi}(b^{\chi*})} \frac{r}{\mu\theta} > \frac{rc_b}{\mu\theta}.$$

In addition, by the convexity of V^{χ} and $V^{\chi'}(b^{\chi*}) = 1$, it follows that

$$\frac{rc_b}{\mu\theta} < V^{\chi'}(0) < V^{\chi'}(b^{\chi*}) = 1,$$

which implies $\mu\theta > rc_b$ and hence $y_b > 0$. By simply comparing the definitions of y_b and y_s, it is clear that $y_b < y_s$. Therefore, by observing that $(\mathcal{L}^{\chi} - r)\hat{h}^{\chi}(y_b) = 0$, we conclude $(\mathcal{L}^{\chi} - r)\hat{h}^{\chi}(y) < 0$ if $y \in [0, y_b)$, and $(\mathcal{L}^{\chi} - r)\hat{h}^{\chi}(y) > 0$ if $y \in (y_b, +\infty)$. This suggests the concavity and convexity of $\hat{H}^{\chi}$ as desired.

Chapter 5

Futures Trading Under Mean Reversion

Futures are an integral part of the universe of derivatives. A futures is a contract that requires the buyer to purchase (seller to sell) a fixed quantity of an asset, such as a commodity, at a fixed price to be paid for on a pre-specified future date. Commonly traded on exchanges, there are futures written on various underlying assets or references, including commodities, interest rates, equity indices, and volatility indices. Many futures stipulate physical delivery of the underlying asset, with notable examples of agricultural, energy, and metal futures. However, some, like the VIX futures, are settled in cash.

In this chapter, we discuss the pricing and trading of futures under mean-reverting spot price dynamics. Our objectives are to investigate the effect of mean reversion on the return characteristics of futures and futures portfolio, and develop dynamic speculative trading strategies. In addition, we will apply our analytical methods to study VIX futures based exchange-traded notes.

5.1 Futures Prices Under Mean-Reverting Spot Models

Throughout this chapter, we consider futures that are written on an asset with mean-reverting price dynamics, and denote S as the spot price. In this section, we discuss the pricing of futures and their term structures under different market environments.

5.1.1 *OU and CIR Spot Models*

We begin with two mean-reverting models for the spot price S, namely, the OU and CIR models. As we will see, they yield the same price function for the futures contract.

To start, suppose that the spot price evolves according to the OU model:

$$dS_t = \mu(\theta - S_t)dt + \sigma dB_t,$$

where $\mu, \sigma > 0$, $\theta \in \mathbb{R}$, and B is a standard Brownian motion under the historical measure $\mathbb{P}$.

To price a futures contract, we assume that the risk-neutral dynamics of S is of the same mean-reverting class as that of S under $\mathbb{P}$. Hence, under measure $\mathbb{Q}$ the spot price evolves according to

$$dS_t = \tilde{\mu}(\tilde{\theta} - S_t)\,dt + \sigma\,dB_t^{\mathbb{Q}},$$

where $B^{\mathbb{Q}}$ is a standard Brownian motion under $\mathbb{Q}$, with constant parameters $\tilde{\mu}, \sigma > 0$, and $\tilde{\theta} \in \mathbb{R}$. This is again an OU process, albeit with a different long-run mean and speed of mean reversion. This involves a change of measure that connects the two Brownian motions, as described by

$$dB_t^{\mathbb{Q}} = dB_t + \frac{\mu(\theta - S_t) - \tilde{\mu}(\tilde{\theta} - S_t)}{\sigma}dt.$$

In this chapter, futures prices are computed the same as forward prices, and we do not distinguish between the two prices; see Cox *et al.* (1981); Brennan and Schwartz (1990). As such, the price of a futures contract with maturity T is given by

$$f_t^T \equiv f(t, S_t; T) := \mathbb{E}^{\mathbb{Q}}\{S_T | S_t\} = (S_t - \tilde{\theta})e^{-\tilde{\mu}(T-t)} + \tilde{\theta}, \qquad (5.1)$$

for $t \leq T$. Note that the futures price is a deterministic function of time and the current spot price.

We now consider the CIR model for the spot price:

$$dS_t = \mu(\theta - S_t)dt + \sigma\sqrt{S_t}dB_t, \qquad (5.2)$$

where $\mu, \theta, \sigma > 0$, and B is a standard Brownian motion under the historical measure $\mathbb{P}$. Under risk-neutral measure $\mathbb{Q}$,

$$dS_t = \tilde{\mu}(\tilde{\theta} - S_t)dt + \sigma\sqrt{S_t}dB_t^{\mathbb{Q}}, \qquad (5.3)$$

where $\mu, \theta > 0$, and $B^{\mathbb{Q}}$ is a $\mathbb{Q}$-standard Brownian motion. In both SDEs, (5.2) and (5.3), we require $2\mu\theta \geq \sigma^2$ and $2\tilde{\mu}\tilde{\theta} \geq \sigma^2$ (Feller condition) so that the CIR process stays positive.

The two Brownian motions are related by

$$dB_t^{\mathbb{Q}} = dB_t + \lambda(S_t)dt,$$

where

$$\lambda(S_t) = \frac{\mu(\theta - S_t) - \tilde{\mu}(\tilde{\theta} - S_t)}{\sigma\sqrt{S_t}}.$$

is the market price of risk. This form of risk premium preserves the CIR model, up to different parameter values across two measures.

The CIR terminal spot price S_T admits the non-central Chi-squared distribution and is positive, whereas the OU spot price is normally distributed, and thus can be positive or negative. Nevertheless, the futures price under the CIR model admits the same functional form as in the OU case (see (5.1)):

$$f_t^T = (S_t - \tilde{\theta})e^{-\tilde{\mu}(T-t)} + \tilde{\theta}, \quad t \le T. \tag{5.4}$$

This allows us to view the spot price as a function of the futures price:

$$S_t = (f_t^T - \tilde{\theta})e^{\tilde{\mu}(T-t)} + \tilde{\theta}.$$

To understand the property of the futures prices, we differentiate (5.4) with respect to T to get

$$\frac{\partial f_t^T}{\partial T} = -\tilde{\mu}(S_t - \tilde{\theta})e^{-\tilde{\mu}(T-t)}. \tag{5.5}$$

The derivative (in T) is positive (resp. negative) if and only if $S_t < \tilde{\theta}$ (resp. $S_t > \tilde{\theta}$). This implies that f_t^T is strictly increasing (resp. decreasing) in T if and only if $S_t < \tilde{\theta}$ (resp. $S_t > \tilde{\theta}$).

A second differentiation of (5.5) with respect to T yields

$$\frac{\partial^2 f_t^T}{\partial T^2} = \tilde{\mu}^2(S_t - \tilde{\theta})e^{-\tilde{\mu}(T-t)}.$$

Hence, the term structure of futures contracts is upward-sloping and concave if $S_t < \tilde{\theta}$. On the other hand, if $S_t > \tilde{\theta}$, then the term structure is downward-slopping and convex. Both term structures are observed empirically (see Figure 5.1 below).

Remark 5.1. The futures price formula (5.4) holds more generally for other mean-reverting models with risk-neutral spot dynamics of the form:

$$dS_t = \tilde{\mu}(\tilde{\theta} - S_t)dt + \sigma(S_t)dB_t^{\mathbb{Q}},$$

where $\sigma(s)$ is a deterministic function of the spot price s such that $\mathbb{E}^{\mathbb{Q}}\{\int_0^T \sigma(S_t)^2 dt\} < \infty$. The OU and CIR models belong to this framework.

5.1.2   *Exponential OU Spot Model*

Under the exponential OU (XOU) model, the spot price follows the SDE:

$$dS_t = \mu(\theta - \ln(S_t))S_t dt + \sigma S_t dB_t, \tag{5.6}$$

with positive parameters (μ, θ, σ), and standard Brownian motion B under the historical measure $\mathbb{P}$. For pricing futures, we assume that the risk-neutral dynamics of S satisfies

$$dS_t = \tilde{\mu}(\tilde{\theta} - \ln(S_t))S_t dt + \sigma S_t dB_t^{\mathbb{Q}},$$

where $\tilde{\mu}, \tilde{\theta} > 0$, and $B^{\mathbb{Q}}$ is a standard Brownian motion under the risk-neutral measure $\mathbb{Q}$.

For a futures contract written on S with maturity T, its price at time t is given by

$$f_t^T = \exp\left(e^{-\tilde{\mu}(T-t)} \ln(S_t) + (1 - e^{-\tilde{\mu}(T-t)})(\tilde{\theta} - \frac{\sigma^2}{2\tilde{\mu}}) \right.$$
$$\left. + \frac{\sigma^2}{4\tilde{\mu}}(1 - e^{-2\tilde{\mu}(T-t)}) \right). \tag{5.7}$$

In turn, we can express the spot price in terms of the futures price:

$$S_t = \exp\left(e^{\tilde{\mu}(T-t)} \ln(f_t^T) + (1 - e^{\tilde{\mu}(T-t)})(\tilde{\theta} - \frac{\sigma^2}{2\tilde{\mu}}) \right.$$
$$\left. + \frac{\sigma^2}{4\tilde{\mu}}(e^{-\tilde{\mu}(T-t)} - e^{\tilde{\mu}(T-t)}) \right).$$

Direct differentiation of f_t^T yields that

$$\frac{\partial f_t^T}{\partial T} = \left[\tilde{\mu}\left(\tilde{\theta} - \frac{\sigma^2}{2\tilde{\mu}} - \ln S_t \right) e^{-\tilde{\mu}(T-t)} + \frac{\sigma^2}{2}e^{-2\tilde{\mu}(T-t)} \right] f_t^T. \tag{5.8}$$

Thus, f_t^T is strictly increasing (resp. decreasing) in T if and only if the spot price is sufficiently low (resp. high):

$$\ln S_t \lessgtr \tilde{\theta} - \frac{\sigma^2}{2\tilde{\mu}}(1 - e^{-\tilde{\mu}(T-t)}).$$

Further differentiation of (5.8) gives

$$\frac{\partial^2 f_t^T}{\partial T^2} = \left[\tilde{\mu}^2 e^{-2\tilde{\mu}(T-t)} \left(\tilde{\theta} - \frac{\sigma^2}{2\tilde{\mu}} - \ln S_t \right)^2 + \frac{\sigma^4}{4}e^{-4\tilde{\mu}(T-t)} - \sigma^2 \tilde{\mu} e^{-2\tilde{\mu}(T-t)} \right.$$
$$\left. + \left(\tilde{\mu}\sigma^2 e^{-3\tilde{\mu}(T-t)} - \tilde{\mu}^2 e^{-\tilde{\mu}(T-t)} \right) \left(\tilde{\theta} - \frac{\sigma^2}{2\tilde{\mu}} - \ln S_t \right) \right] f_t^T.$$

Inspecting this derivative, we arrive at the following scenarios:

(i) The term structure is downward-sloping and convex if

$$\ln S_t > \tilde{\theta} - \frac{\sigma^2}{2\tilde{\mu}}(1 - e^{-\tilde{\mu}(T-t)}) + \left(\frac{e^{2\tilde{\mu}(T-t)}}{4} + \frac{\sigma^2}{2\tilde{\mu}}\right)^{\frac{1}{2}} - \frac{e^{\tilde{\mu}(T-t)}}{2}.$$

(ii) The term structure is downward-sloping and concave if

$$\tilde{\theta} - \frac{\sigma^2}{2\tilde{\mu}}(1 - e^{-\tilde{\mu}(T-t)}) < \ln S_t$$

$$< \tilde{\theta} - \frac{\sigma^2}{2\tilde{\mu}}(1 - e^{-\tilde{\mu}(T-t)}) + \left(\frac{e^{2\tilde{\mu}(T-t)}}{4} + \frac{\sigma^2}{2\tilde{\mu}}\right)^{\frac{1}{2}} - \frac{e^{\tilde{\mu}(T-t)}}{2}.$$

(iii) The term structure is upward-sloping and concave if

$$\tilde{\theta} - \frac{\sigma^2}{2\tilde{\mu}}(1 - e^{-\tilde{\mu}(T-t)}) - \left(\frac{e^{2\tilde{\mu}(T-t)}}{4} + \frac{\sigma^2}{2\tilde{\mu}}\right)^{\frac{1}{2}} - \frac{e^{\tilde{\mu}(T-t)}}{2} < \ln S_t$$

$$< \tilde{\theta} - \frac{\sigma^2}{2\tilde{\mu}}(1 - e^{-\tilde{\mu}(T-t)}).$$

(iv) The term structure is upward-sloping and convex if

$$\ln S_t < \tilde{\theta} - \frac{\sigma^2}{2\tilde{\mu}}(1 - e^{-\tilde{\mu}(T-t)}) - \left(\frac{e^{2\tilde{\mu}(T-t)}}{4} + \frac{\sigma^2}{2\tilde{\mu}}\right)^{\frac{1}{2}} - \frac{e^{\tilde{\mu}(T-t)}}{2}.$$

Figure 5.1 displays two characteristically different term structures observed in the VIX futures market. These futures, written on the CBOE Volatility Index (VIX) are traded on the CBOE Futures Exchange. As the VIX measures the 1-month implied volatility calculated from the prices of S&P 500 options, VIX futures provide exposure to the market's volatility. We plot the VIX futures prices during the recent financial crisis on November 20, 2008 (left), and on a post-crisis date, July 22, 2015 (right), along with the calibrated futures curves under the OU/CIR model and XOU model. In the calibration, the model parameter values are chosen to minimize the sum of squared errors between the model and observed futures prices.

The OU/CIR/XOU model generates a decreasing convex curve for November 20, 2008 (left), and an increasing concave curve for July 22, 2015 (right), and they all fit the observed futures prices very well. The former term structure starts with a very high spot price of 80.86 with a calibrated risk-neutral long-run mean $\tilde{\theta} = 40.36$ under the OU/CIR model, suggesting that the market's expectation of falling market volaitility. In contrast, we infer from the term structure on July 25, 2015 that the market expects the VIX to raise from the current spot value of 12.12 to be closer to $\tilde{\theta} = 18.16$.

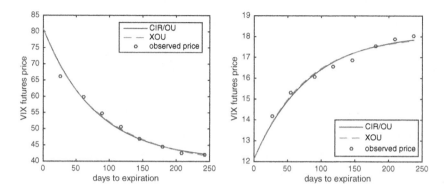

Fig. 5.1   (Left) VIX futures historical prices on Nov 20, 2008 with the current VIX value at 80.86. The days to expiration range from 26 to 243 days (Dec–Jul contracts). Calibrated parameters: $\tilde{\mu} = 4.59, \tilde{\theta} = 40.36$ under the CIR/OU model, or $\tilde{\mu} = 3.25, \tilde{\theta} = 3.65, \sigma = 0.15$ under the XOU model. (Right) VIX futures historical prices on Jul 22, 2015 with the current VIX value at 12.12. The days to expiration ranges from 27 days to 237 days (Aug–Mar contracts). Calibrated parameters: $\tilde{\mu} = 4.55, \tilde{\theta} = 18.16$ under the CIR/OU model, or $\tilde{\mu} = 4.08, \tilde{\theta} = 3.06, \sigma = 1.63$ under the XOU model.

5.2   Roll Yield

By design, the value of a futures contract converges to the spot price as time approaches maturity. If the futures market is in *backwardation*, the futures price increases to reach the spot price at expiry. In contrast, when the market is in contango, the futures price tends to decrease to the spot price. For an investor with a long futures position, the return is positive in a backwardation market, and negative in a contango market. An investor can long the front-month contract, then sell it at or before expiry, and simultaneously go long the next-month contract. This *rolling strategy* that involves repeatedly rolling an expiring contract into a new one is commonly adopted during backwardation, while its opposite is often used in a contango market. Backwardation and contango phenomena are widely observed in the energy commodities and volatility futures markets.

As we can see, both the futures and spot prices vary over time. If the spot price increases/decreases, the futures price will also end up higher/lower. This leads us to consider the difference between the futures and spot returns.[1] Let $0 \leq t_1 < t_2 \leq T$. We denote the *roll yield* over the

---

[1] Here, "return" stands for "the change in the value of an instrument or index between two points in time, without dividing by the initial value" according to *Deconstructing Futures Returns: The Role of Roll Yield*, Campbell White Paper Series, February 2014.

period $[t_1, t_2]$ associated with a single futures contract with maturity T by

$$\mathcal{R}(t_1, t_2, T) := (f_{t_2}^T - f_{t_1}^T) - (S_{t_2} - S_{t_1}). \tag{5.9}$$

If $t_2 = T$, then the roll yield reduces to the price difference $(S_{t_1} - f_{t_1}^T)$.

Next, we examine the cumulative roll yield across maturities. Denote by $T_1 < T_2 < T_3 < \dots$ the maturities of futures contracts. We roll over at every T_i by replacing the contract expiring at T_i with a new contract that expires at T_{i+1}. Let $i(t) := \min\{i : T_{i-1} < t \le T_i\}$, and $i(0) = 1$. Then the roll yield up to time $t > T_1$ is

$$\mathcal{R}(0, t) = (f_t^{T_{i(t)}} - f_{T_{i(t)-1}}^{T_{i(t)}}) + \sum_{j=2}^{i(t)-1} (S_{T_j} - f_{T_{j-1}}^{T_j}) + (S_{T_1} - f_0^{T_1}) - (S_t - S_0)$$

$$= \underbrace{(f_t^{T_{i(t)}} - S_t) - (f_0^{T_1} - S_0)}_{\text{Basis Return}} + \underbrace{\sum_{j=1}^{i(t)-1} (S_{T_j} - f_{T_j}^{T_{j+1}})}_{\text{Cumulative Roll Adjustment}} \quad .$$

The cumulative roll adjustment is related to the term structure of futures contracts. If $T_i - T_{i-1}$ is constant, and the term structure only moves parallel, then the cumulative roll adjustment is simply the number of roll-over times a constant (difference between spot and near-month futures contract).

5.2.1 *OU and CIR Spot Models*

Suppose the spot price under the OU or CIR model described in Section 5.1.1. At time T_1, we roll over the portfolio by selling the futures contracts with maturity T_1 and buying the futures contracts with maturity $T_2 > T_1$. Then, by the futures price formula (5.1), the conditional expected roll yield is

$$\mathbb{E}\{\mathcal{R}(t_1, t_2, T)|S_{t_1}\} = S_{t_1}\left((1 - e^{-\tilde{\mu}(T-t_1)}) - e^{-\mu(t_2-t_1)}(1 - e^{-\tilde{\mu}(T-t_2)})\right)$$

$$- \theta(1 - e^{-\mu(t_2-t_1)})(1 - e^{-\tilde{\mu}(T-t_2)})$$

$$- \tilde{\theta}(e^{-\tilde{\mu}(T-t_2)} - e^{-\tilde{\mu}(T-t_1)}).$$

In particular, if $\tilde{\mu} = \mu$, then the conditional expected roll yield simplifies to

$$\mathbb{E}\{\mathcal{R}(t_1, t_2, T)|S_{t_1}\} = \left((S_{t_1} - \theta) + (\theta - \tilde{\theta})e^{-\mu(T-t_2)}\right)\left(1 - e^{-\mu(t_2-t_1)}\right).$$

From this, we see that the roll yield is positive if $S_{t_1} \ge \theta \ge \tilde{\theta}$. If additionally $S_{t_1} = \theta = \tilde{\theta}$, then the conditional expected roll yield vanishes.

Consider a longer horizon with rolling at multiple maturities, the expected roll yield is

$$\mathbb{E}\{\mathcal{R}(0,t)\}$$

$$= \mathbb{E}\{f_t^{T_{i(t)}} - S_t\} - (f_0^{T_1} - S_0) + \sum_{j=1}^{i(t)-1} \mathbb{E}\{S_{T_j} - f_{T_j}^{T_{j+1}}\}$$

$$= ((S_0 - \theta)e^{-\mu t} + \theta - \tilde{\theta})(e^{-\tilde{\mu}(T_{i(t)}-t)} - 1) - (S_0 - \tilde{\theta})(e^{-\tilde{\mu}T_1} - 1)$$

$$+ \sum_{j=1}^{i(t)-1} ((S_0 - \theta)e^{-\mu T_j} + \theta - \tilde{\theta})(1 - e^{-\tilde{\mu}(T_{j+1}-T_j)}).$$

In particular, if the maturities are separated equally, i.e. $T_{j+1}-T_j \equiv \Delta T$ for all j, then we obtain a simplified expression

$$\mathbb{E}\{\mathcal{R}(0,t)\}$$

$$= ((S_0 - \theta)e^{-\mu t} + \theta - \tilde{\theta})(e^{-\tilde{\mu}((i(t)-1)\Delta T + T_1 - t)} - 1)$$

$$+ \left((S_0 - \theta)\frac{1 - e^{-\mu(i(t)-1)\Delta T}}{1 - e^{-\mu \Delta T}} + (i(t) - 1)(\theta - \tilde{\theta})\right)(1 - e^{-\tilde{\mu}\Delta T})$$

$$- (S_0 - \tilde{\theta})(e^{-\tilde{\mu}T_1} - 1).$$

In summary, the expected roll yield depends not only on the risk-neutral parameters $\tilde{\mu}$ and $\tilde{\theta}$, but also their historical counterparts. It vanishes when $S_0 = \theta = \tilde{\theta}$. This is intuitive because if the current spot price is currently at the long-run mean, and the risk-neutral and historical measures coincide, then the spot and futures prices have little tendency to deviate from the long-run mean. Also, notice that neither the futures price nor the roll yield depends on the volatility parameter σ. This is true under the OU/CIR model, but not the exponential OU model, as we discuss next.

5.2.2 *Exponential OU Dynamics*

We now turn to the exponential OU spot price model discussed in Section 5.1.2. Following the futures price formula (5.7) and the roll yield definition

(5.9), we compute explicitly the conditional expected roll yield:

$$
\mathbb{E}\{\mathcal{R}(t_1, t_2, T)|S_{t_1}\}
$$

$$
= \exp\left\{ e^{-\tilde{\mu}(T-t_2)-\mu(t_2-t_1)} \ln(S_{t_1}) + \left(\theta - \frac{\sigma^2}{2\mu}\right)(1 - e^{-\mu(t_2-t_1)})e^{-\tilde{\mu}(T-t_2)} \right.
$$

$$
+ \frac{\sigma^2}{4\mu}e^{-2\tilde{\mu}(T-t_2)}(1 - e^{-2\mu(t_2-t_1)}) + (1 - e^{-\tilde{\mu}(T-t_2)})(\tilde{\theta} - \frac{\sigma^2}{2\tilde{\mu}})
$$

$$
\left. + \frac{\sigma^2}{4\tilde{\mu}}(1 - e^{-2\tilde{\mu}(T-t_2)}) \right\}
$$

$$
- \exp\left\{ e^{-\tilde{\mu}(T-t_1)} \ln(S_{t_1}) + (1 - e^{-\tilde{\mu}(T-t_1)})(\tilde{\theta} - \frac{\sigma^2}{2\tilde{\mu}}) \right.
$$

$$
\left. + \frac{\sigma^2}{4\tilde{\mu}}(1 - e^{-2\tilde{\mu}(T-t_1)}) \right\}
$$

$$
- \exp\left\{ e^{-\mu(t_2-t_1)} \ln(S_{t_1}) + (1 - e^{-\mu(t_2-t_1)})(\theta - \frac{\sigma^2}{2\mu}) \right.
$$

$$
\left. + \frac{\sigma^2}{4\mu}(1 - e^{-2\mu(t_2-t_1)}) \right\} + S_{t_1}.
$$

Rolling over multiple futures contracts, the expected roll yield is

$$
\mathbb{E}\{\mathcal{R}(0, t)\} = Y_1(t) + Y_2(t) - (f_0^{T_1} - S_0), \tag{5.10}
$$

where

$$
Y_1(t) = \mathbb{E}\{f_t^{T_{i(t)}} - S_t\}
$$

$$
= \exp\left(e^{-\tilde{\mu}(T_{i(t)}-t)-\mu t} \ln(S_0) + \left(\theta - \frac{\sigma^2}{2\mu}\right)(1 - e^{-\mu t})e^{-\tilde{\mu}(T_{i(t)}-t)} \right.
$$

$$
+ \frac{\sigma^2}{4\mu}e^{-2\tilde{\mu}(T_{i(t)}-t)}(1 - e^{-2\mu t}) + (1 - e^{-\tilde{\mu}(T_{i(t)}-t)})(\tilde{\theta} - \frac{\sigma^2}{2\tilde{\mu}})
$$

$$
\left. + \frac{\sigma^2}{4\tilde{\mu}}(1 - e^{-2\tilde{\mu}(T_{i(t)}-t)}) \right)
$$

$$
- \exp\left(e^{-\mu t} \ln(S_0) + (1 - e^{-\mu t})(\theta - \frac{\sigma^2}{2\mu}) + \frac{\sigma^2}{4\mu}(1 - e^{-\mu t}) \right),
$$

and

$$Y_2(t) = \sum_{j=1}^{i(t)-1} \mathbb{E}\{S_{T_j} - f_{T_j}^{T_{j+1}}\}$$

$$= \sum_{j=1}^{i(t)-1} \left(\exp \left(e^{-\mu T_j} \ln(S_0) + (1 - e^{-\mu T_j})(\theta - \frac{\sigma^2}{2\mu}) + \frac{\sigma^2}{4\mu}(1 - e^{-\mu T_j}) \right) \right.$$

$$- \exp \left(e^{-\tilde{\mu}(T_{j+1}-T_j) - \mu T_j} \ln(S_0) + \left(\theta - \frac{\sigma^2}{2\mu} \right) (1 - e^{-\mu T_j}) e^{-\tilde{\mu}(T_{j+1}-T_j)} \right.$$

$$+ \frac{\sigma^2}{4\mu} e^{-2\tilde{\mu}(T_{j+1}-T_j)}(1 - e^{-2\mu T_j})$$

$$\left. \left. + (1 - e^{-\tilde{\mu}(T_{j+1}-T_j)})(\tilde{\theta} - \frac{\sigma^2}{2\tilde{\mu}}) + \frac{\sigma^2}{4\tilde{\mu}}(1 - e^{-2\tilde{\mu}(T_{j+1}-T_j)}) \right) \right).$$

The explicit formula (5.10) for the expected roll yield reveals the nontrivial dependence on the volatility parameter σ, as well as the risk-neutral parameters $(\tilde{\mu}, \tilde{\theta})$ and historical parameters (μ, θ). It is useful for instantly predicting the roll yield after calibrating the risk-neutral parameters from the term structure of the futures prices, and estimating the historical parameters from past spot prices.

5.3 Futures Trading Problem

Let us consider the scenario in which an investor has a long position in a futures contract with expiration date T. With a long position in the futures, the investor can hold it till maturity, but can also close the position early by taking an opposite position at the prevailing market price. At maturity, the two opposite positions cancel each other. This motivates us to investigate the best time to close.

If the investor selects to close the long position at time $\tau \leq T$, then she will receive the market value of the futures, denoted by $f(\tau, S_\tau; T)$, minus the transaction cost $c_s \geq 0$. To maximize the expected discounted value, evaluated under the investor's historical probability measure $\mathbb{P}$ with a constant subjective discount rate $r > 0$, the investor solves the optimal stopping problem

$$\mathcal{V}(t, s) = \sup_{\tau \in \mathcal{T}_{t,T}} \mathbb{E}_{t,s}\{e^{-r(\tau-t)}(f(\tau, S_\tau; T) - c_s)\},$$

where $\mathcal{T}_{t,T}$ is the set of all stopping times, with respect to the filtration generated by S, taking values between t and $\hat{T}$, where $\hat{T} \in (0, T]$

is the trading deadline, which can equal but not exceed the futures' maturity. Throughout this chapter, we continue to use the shorthand notation $\mathbb{E}_{t,s}\{\cdot\} \equiv \mathbb{E}\{\cdot|S_t = s\}$ to indicate the expectation taken under the historical probability measure $\mathbb{P}$.

The value function $\mathcal{V}(t,s)$ represents the expected liquidation value associated with the long futures position. Prior to taking the long position in f, the investor, with zero position, can select the optimal timing to start the trade, or not to enter at all. This leads us to analyze the timing option inherent in the trading problem. Precisely, at time $t \leq T$, the investor faces the optimal entry timing problem

$$\mathcal{J}(t,s) = \sup_{\nu \in \mathcal{T}_{t,T}} \mathbb{E}_{t,s}\left\{e^{-r(\nu-t)}(\mathcal{V}(\nu,S_\nu) - (f(\nu,S_\nu;T) + c_b))\right\},$$

where $c_b \geq 0$ is the transaction cost, which may differ from c_s. In other words, the investor seeks to maximize the expected difference between the value function $\mathcal{V}(\nu, S_\nu)$ associated with the long position and the prevailing futures price $f(\nu, S_\nu; T)$. The value function $\mathcal{J}(t,s)$ represents the maximum expected value of the trading opportunity embedded in the futures. We refer this "long to open, short to close" strategy as the *long-short* strategy.

Alternatively, an investor may well choose to short a futures contract with the speculation that the futures price will fall, and then close it out later by establishing a long position.[2] Given an investor who has a unit short position in the futures contract, the objective is to minimize the expected discounted cost to close out this position at/before maturity. The optimal timing strategy is determined from

$$\mathcal{U}(t,s) = \inf_{\tau \in \mathcal{T}_{t,T}} \mathbb{E}_{t,s}\left\{e^{-r(\tau-t)}(f(\tau,S_\tau;T) + c_b)\right\}.$$

As before, if the investor begins with a zero position, then she can decide when to enter the market by solving

$$\mathcal{K}(t,s) = \sup_{\nu \in \mathcal{T}_{t,T}} \mathbb{E}_{t,s}\left\{e^{-r(\nu-t)}((f(\nu,S_\nu;T) - c_s) - \mathcal{U}(\nu,S_\nu))\right\}.$$

We call this "short to open, long to close" strategy as the *short-long* strategy.

When an investor contemplates entering the market, she can either long or short first. Therefore, on top of the timing option, the investor has an

---

[2]By taking a short futures position, the investor is required to sell the underlying spot at maturity at a pre-specified price. In contrast to the short sale of a stock, a short futures does not involve share borrowing or re-purchasing.

additional choice between the long-short and short-long strategies. Hence, the investor solves the market entry timing problem:

$$\mathcal{P}(t,s) = \sup_{\varsigma \in \mathcal{T}_{t,T}} \mathbb{E}_{t,s}\left\{ e^{-r(\varsigma - t)} \max\{\mathcal{A}(\varsigma, S_\varsigma), \mathcal{B}(\varsigma, S_\varsigma)\} \right\}, \qquad (5.11)$$

with two alternative rewards upon entry defined by

$$\mathcal{A}(\varsigma, S_\varsigma) := \mathcal{V}(\varsigma, S_\varsigma) - (f(\varsigma, S_\varsigma; T) + c_b), \quad (long - short),$$
$$\mathcal{B}(\varsigma, S_\varsigma) := (f(\varsigma, S_\varsigma; T) - c_s) - \mathcal{U}(\varsigma, S_\varsigma) \quad (short - long).$$

5.4   Variational Inequalities and Optimal Trading Strategies

In order to solve for the optimal trading strategies, we study the variational inequalities corresponding to the value functions $\mathcal{J}$, $\mathcal{V}$, $\mathcal{U}$, $\mathcal{K}$ and $\mathcal{P}$. To this end, we first define the operators:

$$\mathcal{L}^{(1)}\{\cdot\} := -r \cdot + \frac{\partial \cdot}{\partial t} + \tilde{\mu}(\tilde{\theta} - s)\frac{\partial \cdot}{\partial s} + \frac{\sigma^2}{2}\frac{\partial^2 \cdot}{\partial s^2},$$

$$\mathcal{L}^{(2)}\{\cdot\} := -r \cdot + \frac{\partial \cdot}{\partial t} + \tilde{\mu}(\tilde{\theta} - s)\frac{\partial \cdot}{\partial s} + \frac{\sigma^2 s}{2}\frac{\partial^2 \cdot}{\partial s^2},$$

$$\mathcal{L}^{(3)}\{\cdot\} := -r \cdot + \frac{\partial \cdot}{\partial t} + \tilde{\mu}(\tilde{\theta} - \ln s)s\frac{\partial \cdot}{\partial s} + \frac{\sigma^2 s^2}{2}\frac{\partial^2 \cdot}{\partial s^2},$$

corresponding to, respectively, the OU, CIR, and XOU models.

The optimal exit and entry problems $\mathcal{J}$ and $\mathcal{V}$ associated with the *long-short* strategy are solved from the following pair of variational inequalities:

$$\max\left\{ \mathcal{L}^{(i)}\mathcal{V}(t,s), (f(t,s;T) - c_s) - \mathcal{V}(t,s) \right\} = 0,$$

$$\max\left\{ \mathcal{L}^{(i)}\mathcal{J}(t,s), (\mathcal{V}(t,s) - (f(t,s;T) + c_b)) - \mathcal{J}(t,s) \right\} = 0,$$

for $(t,s) \in [0,T] \times \mathbb{R}$, with $i \in \{1,2,3\}$ representing the OU, CIR, or XOU model respectively.[3] Similarly, the reverse *short-long* strategy can be determined by numerically solving the variational inequalities satisfied by $\mathcal{U}$ and $\mathcal{K}$:

$$\min\left\{ \mathcal{L}^{(i)}\mathcal{U}(t,s), (f(t,s;T) + c_b) - \mathcal{U}(t,s) \right\} = 0,$$

$$\max\left\{ \mathcal{L}^{(i)}\mathcal{K}(t,s), ((f(t,s;T) - c_s) - \mathcal{U}(t,s)) - \mathcal{K}(t,s) \right\} = 0.$$

---

[3]The spot price is positive, thus $s \in \mathbb{R}_+$, under the CIR and XOU models.

As $\mathcal{V}, \mathcal{J}, \mathcal{U}$, and $\mathcal{K}$ are numerically solved, they become the input to the final problem represented by the value function $\mathcal{P}$. To determine the optimal timing to enter the futures market, we solve the variational inequality

$$\max\left\{ \mathcal{L}^{(i)}\mathcal{P}(t,s),\, \max\{\mathcal{A}(t,s),\mathcal{B}(t,s)\} - \mathcal{P}(t,s) \right\} = 0.$$

The optimal timing strategies are described by a series of boundaries representing the time-varying critical spot price at which the investor should establish a long/short futures position. In the "long to open, short to close" trading problem, where the investor pre-commits to taking a long position first, the market entry timing is described by the "$\mathcal{J}$" boundary in Figure 5.2(a). The subsequent timing to exit the market is represented by the "$\mathcal{V}$" boundary in Figure 5.2(a). As we can see, the investor will long the futures when the spot price is low, and short to close the position when the spot price is high, confirming the buy-low-sell-high intuition.

If the investor adopts the *short-long* strategy, by which she will first short a futures and subsequently close out with a long position, then the optimal market entry and exit timing strategies are represented, respectively, by the "$\mathcal{K}$" and "$\mathcal{U}$" boundaries in Figure 5.2(c). The investor will enter the market by shorting a futures when the spot price is sufficiently high (at the "$\mathcal{K}$" boundary), and wait to close it out when the spot price is low. In essence, the boundaries reflect a sell-high-buy-low strategy.

When there are no transaction costs (see Figures 5.2(b) and 5.2(d)), the waiting region shrinks for both strategies. Practically, this means that the investor tends to enter and exit the market earlier, resulting in more rapid trades. This is intuitive as transaction costs discourage trades, especially near expiry.

In the market entry problem represented by $\mathcal{P}(t,s)$ in (5.11), the investor decides at what spot price to open a position. The corresponding timing strategy is illustrated by two boundaries in Figure 5.3(a). The boundary labeled as "$\mathcal{P} = \mathcal{A}$" (resp. "$\mathcal{P} = \mathcal{B}$") indicates the critical spot price (as a function of time) at which the investor enters the market by taking a *long* (resp. *short*) futures position. The area above the "$\mathcal{P} = \mathcal{B}$" boundary is the "short-first" region, whereas the area below the "$\mathcal{P} = \mathcal{A}$" boundary is the "long-first" region. The area between the two boundaries is the region where the investor should wait to enter. The ordering of the regions is intuitive – the investor should long the futures when the spot price is currently low and short it when the spot price is high. As time approaches maturity, the value of entering the market diminishes. The investor will not start a long/short position unless the spot is very low/high close to maturity. Therefore, the waiting region expands significantly nears expiry.

The investor's exit strategy depends on the initial entry position. If the investor enters by taking a long position (at the "$\mathcal{P} = \mathcal{A}$" boundary), then the optimal exit timing to close her position is represented by the upper boundary with label "$\mathcal{V}$" in Figure 5.2(a). If the investor's initial position is short, then the optimal time to close by going long the futures is described by the lower boundary with label "$\mathcal{U}$" in Figure 5.2(c).

Since the value function $\mathcal{P}$ dominates both $\mathcal{J}$ and $\mathcal{K}$ due to the additional flexibility, it is not surprising that the "$\mathcal{P} = \mathcal{A}$" boundary is lower than the "$\mathcal{J}$" boundary, and the "$\mathcal{P} = \mathcal{B}$" boundary is higher than the "$\mathcal{K}$" boundary, as seen in Figure 5.3(b). This means that the embedded timing option to choose between the two strategies ("long to open, short to close" or "short to open, long to close") induces the investor to delay market entry to wait for better prices.

5.5 Dynamic Futures Portfolios

In addition to trading a single futures contract, one can construct a portfolio of futures of different maturities. Let's consider a self-financing portfolio consists of $k \geq 2$ futures with dynamically re-balanced strategies $(w_t^i)_{0 \leq t \leq T_i}$, $i = 1, \ldots, k$. Also, without loss of generality, consider k futures contracts with ordered maturities $T_1 < T_2 < \cdots < T_k$. For $t < T_1$, the portfolio value evolves according to

$$\frac{dV_t}{V_t} = \sum_{i=1}^{k} w_t^i \frac{df_t^{T_i}}{f_t^{T_i}} + r\,dt. \tag{5.12}$$

where $\sum_{i=1}^{k} w_t^i = 1$. In other words, $w_t^i V_t$ is the cash amount invested in futures i (with maturity T_i) at time t.

In general, the strategy w_t can depend on the history of the spot and other sources of randomness. For our analysis and applications herein, it is sufficient to consider Markovian strategies of the form $w_t = w(t, S_t)$ for some deterministic function $w(t, s)$. Note that futures are priced under the risk neutral measure $\mathbb{Q}$, but the investor observes the portfolio value under the historical measure $\mathbb{P}$. The evolution of V in (5.12) holds path-wise under either measure.

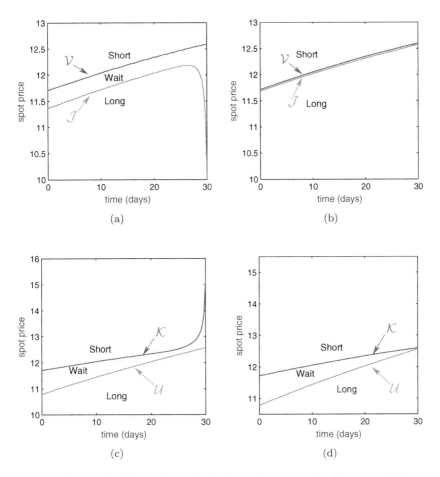

Fig. 5.2   Top: optimal boundaries for the long-short strategies (long at $\mathcal{J}$ first, then close out at $\mathcal{V}$) under the CIR model with and without transaction costs ((a) and (b) respectively).   Bottom: optimal boundaries for the short-long strategies (short at $\mathcal{U}$ first, then close up at $\mathcal{K}$) under the CIR model with and without transaction costs ((c) and (d) respectively).   Parameters: $T = 1/12$ (1 month), $T_f = 3/12$ (3 months), $r = 0.05, \sigma = 0.25, \theta = 17.5, \tilde{\theta} = 18.16, \mu = 5.5, \tilde{\mu} = 4.55, c_b = c_s = 0.0005$.

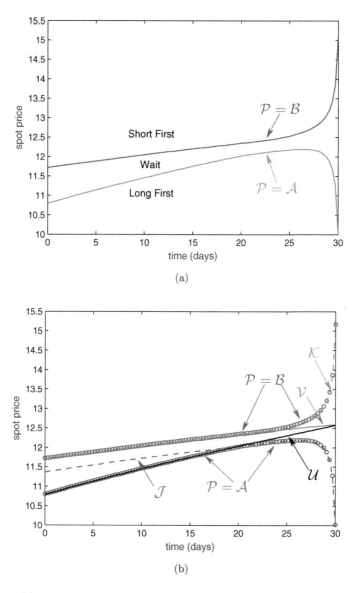

Fig. 5.3 (a) Optimal boundaries to enter the futures market under the CIR model. In the "Short First" (resp. "Long First") region, the investor enters by establishing a short (resp. long) position. (b) Display of all optimal boundaries for entering ($\mathcal{P} = \mathcal{B}, \mathcal{P} = \mathcal{A}, \mathcal{J}, \mathcal{K}$), and subsequently exiting the market ($\mathcal{V}$ and $\mathcal{U}$). Parameters: $T = 1/12$ (1 month), $T_f = 3/12$ (3 months), $r = 0.05, \sigma = 0.25, \theta = 17.5, \tilde{\theta} = 18.16, \mu = 5.5, \tilde{\mu} = 4.55, c_b = c_s = 0.0005$.

In Section 5.1, we have seen how futures prices relate to the spot prices. One major purpose of trading futures is to gain desired exposure to underlying asset. As a simple example, consider a single futures contract under the OU or CIR model. By formula (5.1) and Ito's lemma, the dynamics of this portfolio value is

$$df_t^T = e^{-\tilde{\mu}(T-t)}dS_t + \tilde{\mu}(S_t - \tilde{\theta})e^{-\tilde{\mu}(T-t)}dt.$$

In this equation, the first term suggests that the futures always has a less than a unit exposure to the spot, and the second represents a drift term whose sign and magnitude depend on the relative values of the current spot price and the risk-neutral long-run mean $\tilde{\theta}$.

Now by dynamically trading futures of different maturities, can an investor flexibly *control* the exposure to the spot? Furthermore, we seek to determine the impact of a dynamic trading strategy on the portfolio's rate of return and volatility.

5.5.1 *Portfolio Dynamics with a CIR Spot*

To fix idea, let us discuss the problem under the CIR model. In this case, the futures price is given in (5.1). We apply the futures price formula to (5.12), and derive the dynamics of V in as follows:

$$
\begin{aligned}
\frac{dV_t}{V_t} &= \sum_{i=1}^{k} w_t^i \frac{df_t^{T_i}}{f_t^{T_i}} + rdt \\
&= \sum_{i=1}^{k} w_t^i \frac{\tilde{\mu}(S_t - \tilde{\theta})e^{-\tilde{\mu}(T_i-t)}dt + e^{-\tilde{\mu}(T_i-t)}dS_t}{(S_t - \tilde{\theta})e^{-\tilde{\mu}(T_i-t)} + \tilde{\theta}} + rdt \\
&= \sum_{i=1}^{k} w_t^i \frac{e^{-\tilde{\mu}(T_i-t)}S_t}{(S_t - \tilde{\theta})e^{-\tilde{\mu}(T_i-t)} + \tilde{\theta}} \frac{dS_t}{S_t} \\
&\quad + \sum_{i=1}^{k} w_t^i \frac{\tilde{\mu}(S_t - \tilde{\theta})e^{-\tilde{\mu}(T_i-t)}}{(S_t - \tilde{\theta})e^{-\tilde{\mu}(T_i-t)} + \tilde{\theta}}dt + rdt \\
&= \sum_{i=1}^{k} w_t^i \frac{S_t e^{-\tilde{\mu}(T_i-t)}}{(S_t - \tilde{\theta})e^{-\tilde{\mu}(T_i-t)} + \tilde{\theta}} \frac{dS_t}{S_t} \\
&\quad + \tilde{\mu}(S_t - \tilde{\theta}) \sum_{i=1}^{k} w_t^i \frac{e^{-\tilde{\mu}(T_i-t)}}{(S_t - \tilde{\theta})e^{-\tilde{\mu}(T_i-t)} + \tilde{\theta}}dt + rdt \\
&= \omega(t, S_t) \frac{dS_t}{S_t} + \left(r + \tilde{\mu}(1 - \frac{\tilde{\theta}}{S_t})\omega(t, S_t)\right)dt, \quad (5.13)
\end{aligned}
$$

where we have denoted

$$\omega(t, S_t) := \sum_{i=1}^{k} w_t^i \frac{S_t e^{\tilde{\mu}(T_i - t)}}{(S_t - \tilde{\theta})e^{-\tilde{\mu}(T_i - t)} + \tilde{\theta}}.$$

In (5.13), we have decomposed the portfolio dynamics into two parts: the exposure to the spot and a stochastic drift term. We observe that the magnitude of exposure is represented by the coefficient $\omega(t, S_t)$. In particular, when $\omega(t, S_t) > 1$, the portfolio is said to have a leveraged exposure to the spot. An inverse leverage is achieved when $\omega(t, S_t) < 1$. The investor has full control of the leverage coefficient ω by selecting the portfolio weights w_i for $i = 1, \ldots, k$. Once $\omega(t, s)$ is chosen, the stochastic drift $(r + \tilde{\mu}(1 - \tilde{\theta}/S_t)\omega(t, S_t))$ is simultaneously set.

As an example, let's select $\omega(t, S_t) = \beta$. Consequently, the futures portfolio has a constant leverage ratio β with respect to the spot.

$$\frac{dV_t}{V_t} = \beta \frac{dS_t}{S_t} + \left(r + \beta \tilde{\mu}(1 - \frac{\tilde{\theta}}{S_t})\right) dt.$$

In particular, if $\beta = 1$, then the portfolio has a one-to-one exposure to the spot, but is subject to the stochastic drift rate: $r + \tilde{\mu}(1 - \frac{\tilde{\theta}}{S_t})$, which can be positive or negative depending on the relative values of S and the long-run mean $\tilde{\theta}$. If $\beta = 0$, then we obtain a constant drift rdt, which recovers the risk-free rate.

Moreover, we can express the value of the futures portfolio in terms of the spot price.

Proposition 5.2. *Under the CIR model, the value of a futures portfolio with a constant leverage ($\beta \in \mathbb{R}$) satisfies*

$$\frac{V_t}{V_0} = \left(\frac{S_t}{S_0}\right)^{\beta} \exp\left((r + \beta\tilde{\mu})t + \beta\left(\frac{1}{2}\sigma^2(1 - \beta) - \tilde{\mu}\tilde{\theta}\right)\int_0^t \frac{1}{S_u} du\right).$$

$$(5.14)$$

Proof. Using Ito's formula and setting $\omega(t, S_t) = \beta$ in SDE (5.13), we have

$$d\ln(V_t) = \frac{dV_t}{V_t} - \frac{1}{2}\left(\frac{dV_t}{V_t}\right)^2$$

$$= \beta \frac{dS_t}{S_t} + \left(r + \beta\tilde{\mu}(1 - \frac{\tilde{\theta}}{S_t})\right)dt - \frac{1}{2}\beta^2\left(\frac{dS_t}{S_t}\right)^2$$

$$= \beta d\ln(S_t) + \left[\beta(1 - \beta)\frac{\sigma^2}{2S_t} + \left(r + \beta\tilde{\mu}(1 - \frac{\tilde{\theta}}{S_t})\right)\right]dt. \quad (5.15)$$

Integrating (5.15) gives (5.14). $\qquad\square$

From (5.14), we see that the return of the futures portfolio admits a multiplicative decomposition in terms of two functions of the spot price process. The first term is intuitive as it indicates that the log-return of the futures portfolio is proportional to β times the log-return of the spot, that is, $\ln(V_t/V_0) = \beta \ln(S_t/S_0)$. The second term suggests a stochastic factor that can increase or decrease the portfolio's value. In (5.15), the sign of the drift term depends crucially on the values of β and the ratio $\tilde{\theta}/S_t$. For example, the factor $\beta(1 - \beta)$ is negative if $\beta \notin [0,1]$, and $(1 - \tilde{\theta}/S_t) < 0$ whenever $S_t < \tilde{\theta}$.

Alternatively, one can control the drift, and obtain a stochastic leverage. For example, setting the coefficient of dt to be $a \in \mathbb{R}$, then the portfolio value satisfies

$$\frac{dV_t}{V_t} = \frac{S_t}{\tilde{\mu}(S_t - \tilde{\theta})}(a - r)\frac{dS_t}{S_t} + a\,dt.$$

As a special case, if $a = 0$, then we have

$$\frac{dV_t}{V_t} = -\frac{r}{\tilde{\mu}(1 - \frac{\tilde{\theta}}{S_t})}\frac{dS_t}{S_t}.$$

The resulting leverage ratio is stochastic, and is negative if $S_t > \tilde{\theta}$, and positive if $S_t < \tilde{\theta}$.

5.5.2 *Portfolio Dynamics with an XOU Spot*

We now look at the portfolio dynamics under the exponential OU model. With the spot price given in (5.6) and futures price in (5.7), the portfolio value satisfies

$$
\begin{aligned}
\frac{dV_t}{V_t} &= \sum_{i=1}^{k} w_t^i \frac{df_t^{T_i}}{f_t^{T_i}} + r\,dt \\
&= \sum_{i=1}^{k} w_t^i e^{-\tilde{\mu}(T_i - t)}\frac{dS_t}{S_t} - \tilde{\mu}(\tilde{\theta} - \ln(S_t))\sum_{i=1}^{k} w_t^i e^{-\tilde{\mu}(T_i - t)}dt + r\,dt \\
&= \varpi(t, S_t)\frac{dS_t}{S_t} + (r + \tilde{\mu}(\ln(S_t) - \tilde{\theta})\varpi(t, S_t))dt,
\end{aligned}
\tag{5.16}
$$

where

$$\varpi(t, S_t) := \sum_{i=1}^{k} w_t^i e^{-\tilde{\mu}(T_i - t)}.$$

In (5.16), the coefficient $\varpi(t, S_t)$ represents the stochastic leverage of the portfolio with respect to the spot. In particular, let $\varpi(t, S_t)$ be a constant

$\beta \in \mathbb{R}$. Then, the drift must be $r + \beta\tilde{\mu}(\ln(S_t) - \tilde{\theta})$, which is positive when S_t is sufficiently high (resp. low) for $\beta > 0$ (resp. $\beta < 0$):

$$S_t \gtrless \exp\left(\tilde{\theta} - \frac{r}{\beta\tilde{\mu}}\right), \qquad \text{for } \beta \gtrless 0.$$

Furthermore, we can express the futures portfolio value in terms of the spot price explicitly.

Proposition 5.3. *Under the exponential OU model, the value of a futures portfolio with a constant leverage ($\beta \in \mathbb{R}$) is given by*

$$\frac{V_t}{V_0} = \left(\frac{S_t}{S_0}\right)^{\beta} \exp\left(\beta\,(1-\beta)\,\frac{\sigma^2}{2}t + \beta\tilde{\mu}\left[\int_0^t (\ln(S_u) - \tilde{\theta})du\right] + rt\right).$$

$$(5.17)$$

Proof. We apply Ito's formula and set the leverage $\varpi(t, S_t) = \beta$ in (5.16) to get

$$d\ln(V_t) = \frac{dV_t}{V_t} - \frac{1}{2}\left(\frac{dV_t}{V_t}\right)^2$$

$$= \beta d\ln(S_t) + \beta\left(\frac{\sigma^2}{2}(1-\beta) - \tilde{\mu}(\tilde{\theta} - \ln(S_t))\right)dt + rdt. \quad (5.18)$$

Integrating (5.18) yields (5.17). $\qquad\qquad\square$

Equivalently, we can write (5.17) in log-returns:

$$\ln\left(\frac{V_t}{V_0}\right) = \beta\ln\left(\frac{S_t}{S_0}\right) + \beta\,(1-\beta)\,\frac{\sigma^2}{2}t + \beta\tilde{\mu}\left[\int_0^t (\ln(S_u) - \tilde{\theta})du\right] + rt.$$

From this, the portfolio's log-return can be decomposed into two parts: β times the spot's log-return, and a stochastic rate whose sign depends on the leverage factor and the difference of log-price of the spot and its log-long-run mean. In particular, if $\beta \notin [0,1]$, then the term $\beta(1-\beta)0.5\sigma^2 t$ reduces the portfolio value, and this erosion is worse if the spot has a higher volatility.

On the other hand, if we fix the drift of (5.16) to be a constant $a \in \mathbb{R}$, then the leverage coefficient must be $(a-r)[\tilde{\mu}(\ln(S_t) - \tilde{\theta})]^{-1}$. In the case with $a = 0$, the portfolio value satisfies

$$\frac{dV_t}{V_t} = \frac{r}{\tilde{\mu}(\tilde{\theta} - \ln(S_t))}\frac{dS_t}{S_t}.$$

Hence, the portfolio has a specific stochastic leverage that changes sign depending on the relative values of $\tilde{\theta}$ and S_t. In particular, a one-to-one exposure without drift is not possible with this futures portfolio.

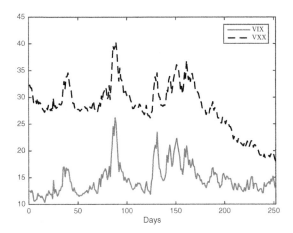

Fig. 5.4 Historical price paths of CBOE Volatility Index (VIX) and iPath S&P 500 VIX Short Term Futures ETN (VXX), respectively, from June 12, 2014 to June 11, 2015.

5.6 Application to VIX Futures & Exchange-Traded Notes

A number of VIX ETFs/ETNs are in fact VIX futures portfolios with time-deterministic weights. For instance, the iPath S&P 500 VIX Futures ETN (VXX) is one of the most traded ETNs. Figure 5.4 displays the time series of the VXX, along with the VIX.

According to its prospectus, the VXX portfolio weights are of the following form:

$$w(t) = \frac{T_{i(t)} - t}{T_{i(t)} - T_{i(t)-1}}, \tag{5.19}$$

where $i(t) := \min\{i : T_{i-1} < t \leq T_i\}$. When the front-month futures expires, the portfolio assigns all weight to the next-month futures. Figure 5.5 illustrates how the weight for each futures contract gradually decreases from 1 to 0 over time.

Under the OU or CIR model, the portfolio dynamics can be described by the SDE:

$$\frac{dV_t}{V_t} = w(t)\frac{df_t^{T_{i(t)}}}{f_t^{T_{i(t)}}} + (1 - w(t))\frac{df_t^{T_{i(t)+1}}}{f_t^{T_{i(t)+1}}} + r\,dt$$

$$= \widehat{\omega}(t, S_t)\frac{dS_t}{S_t} + \left(r + \tilde{\mu}(1 - \frac{\tilde{\theta}}{S_t})\widehat{\omega}(t, S_t)\right)dt. \tag{5.20}$$

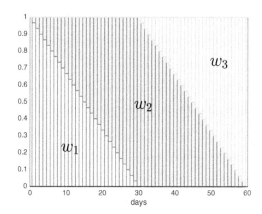

Fig. 5.5 Time-deterministic portfolio weights, as shown in (5.19), for three futures contracts with 30, 60, and 90 days to maturities (respectively w_1, w_2, and w_3).

In (5.20), we have expressed the exposure to the spot in terms of

$$\widehat{\omega}(t, S_t) := \frac{w(t)S_t}{S_t + \tilde{\theta}(e^{\tilde{\mu}(T_{i(t)}-t)} - 1)} + \frac{(1 - w(t))S_t}{S_t + \tilde{\theta}(e^{\tilde{\mu}(T_{i(t)+1}-t)} - 1)}. \quad (5.21)$$

Note that although the weights $w(t)$ is a time-deterministic strategy, the resulting portfolio V has a stochastic leverage with respect to the spot S. We can see it from $\widehat{\omega}$ which is a function of the current spot price as well as time. By inspecting (5.21), we get

$$0 \le \widehat{\omega}(t, S_t) \le w(t) + (1 - w(t)) = 1.$$

Thus, the stochastic exposure always long but is less than 1.

   This portfolio only consists of long positions in the VIX futures, replenished repeatedly over time. Thus its value may erode when persistent negative roll yields are observed in the VIX futures market.

   In Figure 5.6, we show the simulated paths of the futures portfolio, along with the spot and futures prices. The spot price path is more volatile than the VIX futures and the futures portfolio. In fact, we can see this through the quadratic variation of the portfolio return:

$$\left(\frac{dV_t}{V_t}\right)^2 = \left(\frac{w(t)}{S_t + \tilde{\theta}(e^{\tilde{\mu}(T_{i(t)}-t)} - 1)} + \frac{(1 - w(t))}{S_t + \tilde{\theta}(e^{\tilde{\mu}(T_{i(t)+1}-t)} - 1)}\right)^2 (dS_t)^2$$

$$< \left(\frac{dS_t}{S_t}\right)^2.$$

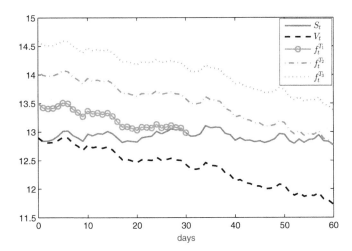

Fig. 5.6 Simulated spot price path (solid line) under CIR model, along with the prices of three futures contracts with maturities $T_1 = 30$ days (circles), $T_2 = 60$ days (dash-dot line), and $T_3 = 90$ days (dotted line), respectively, and a futures portfolio (dashed line) with weights in Figure 5.5. Parameters used for illustration are $\mu = 0.24$, $\theta = 12.90$, $\sigma = 0.29$, $\tilde{\mu} = 0.83$, $\tilde{\theta} = 22.04$, and we set $S_0 = V_0 = 12.90$.

We note that the above inequality holds regardless of the model parameters, and thus, the roll yields of the associated futures.

The simulated VIX index in Figure 5.6 is mean-reverting, with its value staying close to 13. However, the futures prices and portfolio value decrease significantly over time. This is mainly due to the fact that the futures are priced with a high long-run mean $\tilde{\theta}$ while the historical long-run mean θ. Consequently, the futures prices are higher than the spot price. As time progresses, the futures prices must decrease to meet the spot price, so the value of the long-futures portfolio also decays. Such a value erosion is also observed empirically in the VXX prices (see Figure 5.4).

Chapter 6

Optimal Liquidation of Options

For decades, options have been widely used as a tool for investment and risk management. As of 2012, the daily market notional for S&P 500 options is about US$90 billion and the average daily volume has grown rapidly from 119,808 in 2002 to 839,108 as of Jan 2013.[1] Empirical studies on options returns often assume that the options are held to maturity. For every liquidly traded option, there is an embedded timing flexibility to liquidate the position through the market prior to expiry. Hence, an important question for effective risk management is: When is the best time to sell an option? In this chapter, we propose a risk-adjusted optimal timing framework to address this problem for a variety of options under different underlying price dynamics.

In determining the optimal time to sell an option, we incorporate a risk penalty that accounts for adverse price movements till the liquidation time. For every candidate strategy, we measure the associated risk by integrating over time the realized shortfall, or more generally its transformation in terms of a loss function, of the option position. As such, our integrated shortfall risk penalty is path dependent and introduces the trade-off between risk and return for every liquidation timing strategy.

In Section 6.1, we formulate the optimal liquidation problem for a generic European claim in a diffusion market. The investigation of the non-trivial liquidation strategies involves the analytical and numerical studies of the inhomogeneous variational inequality associated with the optimal stopping problem. In Section 6.2, we study the optimal liquidation timing with a shortfall risk penalty. The analysis with a quadratic variation risk penalty is conducted in Section 6.3. For the variational inequalities in our liquidation problems, we prove the existence and uniqueness of a strong solution

---

[1] See http://www.cboe.com/micro/spx/introduction.aspx.

à la Bensoussan and Lions (1978) (see Section 6.5 below) under general conditions applicable to both geometric Brownian motion (GBM) and exponential Ornstein-Uhlenbeck (OU) models for the underlying dynamics. Numerical examples of the optimal liquidation strategies are provided for stocks, calls, puts, and straddles.

6.1   Optimal Liquidation with Risk Penalty

Given a probability space $(\Omega, \mathcal{F}, \mathbb{P})$, where $\mathbb{P}$ is the historical probability measure, we consider a market consisting of a risky asset S and a money market account with a constant positive interest rate r. The risky asset price is modeled by a positive diffusion process following the stochastic differential equation

$$dS_t = \mu(t, S_t)S_t dt + \sigma(t, S_t)S_t dW_t, \qquad S_0 = s, \qquad (6.1)$$

where W is a standard Brownian motion under measure $\mathbb{P}$ and $s > 0$. Here, the deterministic coefficients $\mu(t, s)$ and $\sigma(t, s)$ are assumed to satisfy standard Lipschitz and growth conditions to ensure a unique strong solution to the SDE (see Section 5.2 of Karatzas and Shreve (1991)). We let $\mathbb{F} = (\mathcal{F}_t)_{t \geq 0}$ be the filtration generated by the Brownian motion W.

Let us consider a market-traded European option with payoff $h(S_T)$ on expiration date T written on the underlying asset S. Given that the Sharpe ratio $\lambda(t, s) := \frac{\mu(t,s)-r}{\sigma(t,s)}$ satisfies the Novikov condition: $\mathbb{E}\{\exp(\int_0^T \frac{1}{2}\lambda^2(u, S_u)\, du)\} < \infty$ (see Appendix A of Leung and Shirai (2015)), the density process

$$\left.\frac{d\mathbb{Q}}{d\mathbb{P}}\right|_{\mathcal{F}_t} = \exp\left(-\frac{1}{2}\int_0^t \lambda^2(u, S_u)\, du + \int_0^t \lambda(u, S_u)\, dW_u\right), \qquad (6.2)$$

for $0 \leq t \leq T$, is a $(\mathbb{P}, \mathbb{F})$–martingale. This defines a unique equivalent martingale (risk-neutral) measure $\mathbb{Q}$, and the market price of the option is given by

$$V(t, s) = \widetilde{\mathbb{E}}_{t,s}\left\{e^{-r(T-t)}h(S_T)\right\}, \qquad (t, s) \in [0, T] \times \mathbb{R}^+. \qquad (6.3)$$

The shorthand notation $\widetilde{\mathbb{E}}_{t,s}\{\cdot\} \equiv \widetilde{\mathbb{E}}\{\cdot | S_t = s\}$ denotes the conditional expectation under $\mathbb{Q}$. Note that the market price function $V(t, s)$ does not depend on the drift function $\mu(t, s)$.

Observing the stock and option price movements over time, the investor has the timing flexibility to sell the option before expiry. While seeking to maximize the expected discounted market value of the option, we incorporate a risk penalty that accounts for the downside risk up to the liquidation time. Specifically, we define the shortfall at time t by

$$\ell(t, S_t) = (m - V(t, S_t))^+,$$

where $m > 0$ is a constant benchmark set by the investor. Then, the risk penalty is modeled as a *loss function* of the shortfall, denoted by $\psi(\ell(t, S_t))$. Here, the loss function $\psi : \mathbb{R}^+ \to \mathbb{R}$ is assumed to be increasing, convex, continuously differentiable, with $\psi(0) = 0$ (see Section 4.9 of Föllmer and Schied (2004)). As a result, the investor faces the penalized optimal stopping problem

$$
\begin{aligned}
&J^\alpha(t, s) \\
&= \sup_{\tau \in \mathcal{T}_{t,T}} \mathbb{E}_{t,s} \left\{ e^{-r(\tau-t)} V(\tau, S_\tau) - \alpha \int_t^\tau e^{-r(u-t)} \psi \left((m - V(t, S_t))^+ \right) du \right\},
\end{aligned}
$$
$$(6.4)$$

where $\alpha \geq 0$ is a penalization coefficient and $\mathcal{T}_{t,T}$ is the set of $\mathbb{F}$-stopping times taking values in $[t, T]$.

Unless otherwise noted, our analysis applies to a general loss function ψ satisfying the conditions above. Here, let us give an example to visualize the penalization mechanism. For instance, one can set the benchmark to be the initial option price, and take $\psi(\ell) = \ell$. Then, the penalty term amounts to accumulating the (discounted) area when the option is below its initial cost. We illustrate this in Figure 6.1. Notice that the realized shortfall stays flat when the option price is above the benchmark, and continues to increase as long as the option is under water. Other viable specifications include the power penalty $\psi(\ell) = \ell^p$, $p \geq 1$, and the exponential penalty $\psi(\ell) = \exp(\gamma\ell) - 1$, $\gamma > 0$, and more.

6.1.1 Optimal Liquidation Premium

In order to quantify the value of optimal waiting, we define the optimal liquidation premium by the difference between the value function J^α and the current market price of the option, namely,

$$L^\alpha(t, s) := J^\alpha(t, s) - V(t, s). \tag{6.5}$$

Alternatively, the optimal liquidation premium L^α can be interpreted as the risk-adjusted expected return from a simple buy-now-sell-later strategy.

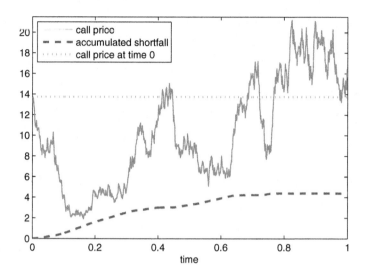

Fig. 6.1   The realized shortfall (dashed) based on a simulated price path (solid) of a European call option under the GBM model, with parameters $S_0 = 100$, $r = 0.03$, $\mu = -0.05$ and $\sigma = 0.3$, $K = 100$, $T = 1$, $\alpha = 1$. The benchmark m is the initial call option price.

Denote the discounted penalized liquidation value process by

$$Y_u = e^{-ru}V(u, S_u) - \alpha \int_0^u e^{-rt}\psi((m - V(t, S_t))^+)dt.$$

In order to guarantee the existence of an optimal stopping time to problem (6.4), we require that $\mathbb{E}\{\sup_{0 \leq u \leq T} Y_u\} < \infty$. For a European call option, the option value $V(t, S_t)$ is dominated by the stock price S_t, while the put option price is bounded by the strike price. Consequently, for any linear combination of calls and puts, it suffices to impose $\mathbb{E}\{\sup_{0 \leq u \leq T} S_u\} < \infty$. We also require that $\mathbb{P}\{\min_{0 \leq t \leq \hat{t}} S_t > 0\} = 1$, which means that the asset price stays strictly positive before any finite time $\hat{t}$ a.s. Then by standard optimal stopping theory, the optimal liquidation time, associated with $L(t, s)$, is given by

$$\tau^* = \inf\{\, u \in [t, T] \,:\, L^\alpha(u, S_u) = 0 \,\}. \tag{6.6}$$

In other words, it is optimal for the investor to sell the option as soon as the optimal liquidation premium L^α vanishes, meaning that the timing flexibility has no value. Accordingly, the investor's optimal liquidation strategy can be described by the sell region $\mathcal{S}$ and delay region $\mathcal{D}$, namely,

$$\mathcal{S} = \{(t, s) \in [0, T] \times \mathbb{R}^+ : L^\alpha(t, s) = 0\},$$
$$\mathcal{D} = \{(t, s) \in [0, T] \times \mathbb{R}^+ : L^\alpha(t, s) > 0\}.$$

Our framework can be readily applied to the *reverse* problem of optimally timing to buy an option. This amounts to changing the sup to inf in L^α. In this chapter, we shall focus on the liquidation problem.

Theorem 6.1. *Given the underlying price dynamics in* (6.1), *the optimal liquidation premium admits the probabilistic representation*

$$L^\alpha(t, s) = \sup_{\tau \in \mathcal{T}_{t,T}} \mathbb{E}_{t,s} \left\{ \int_t^\tau e^{-r(u-t)} G^\alpha(u, S_u) \, du \right\}, \tag{6.7}$$

where we denote

$$G^\alpha(t, s) := \big(\mu(t, s) - r\big) s V_s(t, s) - \alpha \psi \big((m - V(t, s))^+\big). \tag{6.8}$$

Proof. Applying Ito's formula to the market price in (6.3), we get

$$\mathbb{E}_{t,s} \left\{ e^{-r(\tau-t)} V(\tau, S_\tau) \right\} - V(t, s)$$
$$= \mathbb{E}_{t,s} \left\{ \int_t^\tau e^{-r(u-t)} \big(\mu(u, S_u) - r\big) S_u V_s(u, S_u) du \right\}.$$

Substituting this into the optimal liquidation premium in (6.5) gives

$$L^\alpha(t, s) = \sup_{\tau \in \mathcal{T}_{t,T}} \mathbb{E}_{t,s} \left\{ \int_t^\tau e^{-r(u-t)} \big[\big(\mu(u, S_u) - r\big) S_u V_s(u, S_u) \right.$$
$$\left. - \alpha \psi((m - V(u, S_u))^+) \big] \, du \right\},$$

which resembles (6.7). $\qquad\qquad\square$

We shall call $G^\alpha(t, s)$ in (6.8) the *drive function*. We observe that it depends on the Delta $V_s \equiv \frac{\partial V}{\partial s}$ of the option and the penalty coefficient α reduces the drive function for every (t, s). Many properties of the optimal liquidation premium L^α can be deduced by studying the drive function.

Proposition 6.2. *Denote by $t \in [0, T]$ the current time. If the drive function $G^\alpha(u, s)$ is positive, $\forall (u, s) \in [t, T] \times \mathbb{R}^+$, then it is optimal to sell at maturity, namely, $\tau^* = T$. In contrast, if the drive function $G^\alpha(u, s)$ is negative, $\forall (u, s) \in [t, T] \times \mathbb{R}^+$, then it is optimal to sell immediately, namely, $\tau^* = t$.*

Proof. We observe from the integral in (6.7) that if the drive function G^α is positive (resp. negative), $\forall (u, s) \in [t, T] \times \mathbb{R}^+$, then we can maximize the expectation by selecting the largest (resp. smallest) stopping time, namely, $\tau^* = T$ (resp. $\tau^* = t$). $\qquad\qquad\square$

In particular, if $V_s(t, s)$ and $(\mu(t, s) - r)$ are of different signs $\forall (t, s)$, then the drive function G^α is always negative, so it is optimal to sell immediately. Proposition 6.2 can also be applied to the perpetual case if we set $T = \infty$. In general, the delay region always contains the region where the drive function is positive, namely,

$$\{G^\alpha > 0\} \subset \{L^\alpha > 0\}, \tag{6.9}$$

see, for example, Prop. 2.3 of Oksendal and Sulem (2005). Intuitively, this means that if $G(t, s) > 0$, then the investor should not sell immediately since an incremental positive infinitesimal premium can be obtained by waiting for an infinitesimally small amount of time.

In addition, we can infer from (6.7) the ordering of optimal liquidation premium based on the drive function.

Corollary 6.3. *Consider two options A and B, along with two penalty coefficients α_A and α_B respectively. If the drive function of A dominates that of B, i.e. $G_A^{\alpha_A}(t, s) \geq G_B^{\alpha_B}(t, s), \forall (t, s) \in [0, T] \times \mathbb{R}^+$, then the optimal liquidation premium for A, $L_A^{\alpha_A}$, dominates that for B, $L_B^{\alpha_B}(t, s)$, i.e. $L_A^{\alpha_A}(t, s) \geq L_B^{\alpha_B}(t, s), \forall (t, s) \in [0, T] \times \mathbb{R}^+$.*

The corollary allows us to compare the liquidation timing of different penalties. For example, for $0 \leq \alpha_1 \leq \alpha_2$, we have $G^{\alpha_1}(t, s) \geq G^{\alpha_2}(t, s)$ for the same option. It follows from (6.6) and Corollary 6.3 that the optimal liquidation time with penalty α_1 is later than that with penalty α_2.

In general, a variety of delay and sell regions can occur depending on the underlying dynamics and option payoff. Next, we give sufficient conditions so that the delay region is bounded.

Theorem 6.4. *Let $T < \infty$ and S be time homogeneous. Then, the delay region is bounded provided that*
(i) $\exists c > 0$ s.t. $G^\alpha(t, s) < c$ for every $(t, s) \in [0, T] \times \mathbb{R}^+$; and
(ii) there exist constants $b, k > 0$ such that $G^\alpha(t, s) < -b$ in $[0, T] \times [k, \infty)$.

Proof. Step 1. *We find a function $\widehat{L}(t,s)$ that dominates $L^\alpha(t,s)$ and is decreasing in both t and s.* To this end, we define

$$\widehat{G}^\alpha(s) := \max\{G^\alpha(t,\xi) : (t,\xi) \in [0,T] \times [s,\infty)\},$$

$$\widehat{L}(t,s) := \sup_{\tau \in \mathcal{T}_{t,T}} \mathbb{E}_{t,s}\left\{\int_t^\tau e^{-r(u-t)}\widehat{G}^\alpha(S_u)\,du,\right\}.$$

By construction $\widehat{G}^\alpha : [0,T] \times \mathbb{R}^+ \to \mathbb{R}$ is constant in t and decreasing in s. It also satisfies conditions (i) and (ii). Consequently, using the time homogeneity of S, we have, for $t > t'$,

$$\widehat{L}(t,s) = \sup_{\tau \in \mathcal{T}_{0,T-t}} \mathbb{E}_{0,s}\left\{\int_0^\tau e^{-ru}\widehat{G}^\alpha(S_u)\,du\right\}$$

$$\leq \sup_{\tau \in \mathcal{T}_{0,T-t'}} \mathbb{E}_{0,s}\left\{\int_0^\tau e^{-ru}\widehat{G}^\alpha(S_u)\,du\right\} = \widehat{L}(t',s).$$

Hence, $\widehat{L}(t,s)$ is decreasing in t. Moreover, since $\widehat{G}^\alpha$ is decreasing in s, we have, for $s' > s$,

$$\widehat{L}(t,s') = \sup_{\tau \in \mathcal{T}_{t,T}} \mathbb{E}_{t,s'}\left\{\int_t^\tau e^{-r(u-t)}\widehat{G}^\alpha(S_u)\,du\right\}$$

$$= \sup_{\tau \in \mathcal{T}_{t,T}} \mathbb{E}_{t,s}\left\{\int_t^\tau e^{-r(u-t)}\widehat{G}^\alpha(S_u + s' - s)\,du\right\}$$

$$\leq \sup_{\tau \in \mathcal{T}_{t,T}} \mathbb{E}_{t,s}\left\{\int_t^\tau e^{-r(u-t)}\widehat{G}^\alpha(S_u)\,du\right\} = \widehat{L}(t,s).$$

Therefore, $\widehat{L}(t,s)$ is also decreasing in s.

Since by definition $\widehat{G}^\alpha$ dominates G^α, Corollary 6.3 implies that $L^\alpha(t,s)$ has a bounded support as long as $\widehat{L}(t,s)$ has a bounded support. Henceforth, we can assume without loss of generality that L^α is decreasing in both variables t and s and that G^α is time homogeneous and decreasing in s. In particular, we denote $G^\alpha(s) \equiv G^\alpha(t,s)$.

Step 2. *We prove that for every $t > 0$, there exists $\hat{s} < \infty$ such that $L^\alpha(t,s) = 0$ for every $s > \hat{s}$.* Since $L^\alpha(t,s)$ is decreasing, it is equivalent to show that there exists no $\hat{t} \in (0,T]$ s.t. $L^\alpha(t,s) > 0$ for $0 \leq t \leq \hat{t}$ and $s \in \mathbb{R}^+$. To this end, let us suppose that such a time $\hat{t}$ exists. In other words, $\tau^* = \inf\{t \leq u \leq T : L(u, S_u) = 0\} > \hat{t}$.

Now, we show that this leads to a contradiction. Fix $t \in [0,\hat{t})$. Condition (ii) means that there exists k s.t. $G^\alpha(s) < -b < 0$ in $[k,\infty)$. For $s > k$, we let $\tau_k := \inf\{u \geq t : S_u \leq k\}$. Since S has continuous paths, we have

$\tau_k > t$. Define

$$
\mathcal{K}(t, s) := \frac{c}{r} \mathbb{E}_{t,s} \left\{ e^{-r(\tau_k - t)} \mathbf{1}_{\{\tau_k \leq \tau^*\}} \right\} - \mathbb{E}_{t,s} \left\{ b \int_t^{\tau^* \wedge \tau_k} e^{-r(u-t)} \, du \right\}
$$

$$
= \frac{c}{r} \mathbb{E}_{t,s} \left\{ e^{-r(\tau_k - t)} \mathbf{1}_{\{\tau_k \leq \tau^*\}} \right\} - b \left(1 - \mathbb{E}_{t,s} \left\{ e^{-r(\tau^* \wedge \tau_k - t)} \right\} \right),
$$

$$(6.10)$$

where c is the upper bound of G^α in condition (i). Next, taking $s \uparrow \infty$ yields that $\mathbb{P}_{t,s}(\tau_k \leq T) \downarrow 0$, while $\mathbb{E}_{t,s} \left\{ e^{-r(\tau^* \wedge \tau_k - t)} \right\} < e^{-r(\hat{t} - t)}$ since $\tau^* > \hat{t} > t$ a.s. Therefore, we obtain

$$
\beta(t, s) := \frac{c \, \mathbb{P}_{t,s}(\tau_k \leq \tau^*)}{r(1 - \mathbb{E}_{t,s} \left\{ e^{-r(\tau^* \wedge \tau_k - t)} \right\})} \to 0.
$$

As a result, for a sufficiently large $s > k$, we get $b \geq \beta(t, s)$, which implies that $\mathcal{K}(t, s) \leq 0$ (see (6.10)).

Next we consider the difference

$$
L^\alpha(t, s) - \mathcal{K}(t, s) \leq \mathbb{E}_{t,s} \left\{ \int_{\tau^* \wedge \tau_k}^{\tau^*} e^{-r(u-t)} G^\alpha(S_u) \, du - \frac{c}{r} e^{-r(\tau_k - t)} \mathbf{1}_{\{\tau_k \leq \tau^*\}} \right\}
$$

$$
\leq e^{rt} \mathbb{E}_{t,s} \left\{ \frac{c}{r} (e^{-r\tau^* \wedge \tau_k} - e^{-r\tau^*}) - \frac{c}{r} e^{-r\tau_k} \mathbf{1}_{\{\tau^* \leq \tau_k\}} \right\}
$$

$$
= -\frac{ce^{rt}}{r} \mathbb{E}_{t,s} \left\{ e^{-r\tau^*} \mathbf{1}_{\{\tau_k \leq \tau^*\}} \right\} \leq 0.
$$

This means that $L^\alpha(t, s) \leq \mathcal{K}(t, s) \leq 0$. This contradicts the assumption $L^\alpha(t, s) > 0$.

Step 3. It remains to show at time 0 that $\exists \hat{s} > 0$ such that $L^\alpha(0, s) = 0$ for every $s > \hat{s}$. Let $\hat{t} \in [0, T]$ and consider for every $t \in [0, T + \hat{t}]$ the optimal stopping problem

$$
\overline{L}^\alpha(t, s) := \sup_{\tau \in \mathcal{T}_{t, T+\hat{t}}} \mathbb{E}_{t,s} \left\{ \int_t^\tau e^{-r(u-t)} G^\alpha(S_u) \, du \right\}.
$$

The time homogeneity of S yields that $\overline{L}^\alpha(\hat{t}, s) = L^\alpha(0, s)$. Now we apply Step 2 and conclude that there exists $\hat{s} > 0$ such that $\overline{L}^\alpha(\hat{t}, s) = 0$ for every $s > \hat{s}$. Hence, the delay region is bounded above. $\qquad\square$

We remark that the statement and the proof of Theorem 6.4 do not involve the properties of the loss function. In other words, as long as the resulting drive function satisfies conditions (i) and (ii), the delay region is bounded. We notice that if the delay region is bounded, then there exists a constant $\bar{s}$ such that $\{L^\alpha > 0\} \subseteq [0, T] \times (0, \bar{s})$. We will utilize Theorem 6.4 repeatedly when we discuss the liquidation strategies in subsequent sections.

6.2   Applications to GBM and Exponential OU Models

Henceforth, we shall investigate analytically and numerically the optimal liquidation timing when S follows (i) the geometric Brownian motion (GBM) model with $\mu(t,s) = \mu$ and $\sigma(t,s) = \sigma > 0$, as well as (ii) the exponential OU model with $\mu(t,s) = \beta(\theta - \log(s))$ and $\sigma(t,s) = \sigma > 0$.

We will study the liquidation timing of a stock, European put and call options. For both the GBM and exponential OU cases, the risk-neutral measure $\mathbb{Q}$ is uniquely defined by (6.2), and the Novikov condition is satisfied. Furthermore, the $\mathbb{Q}$ dynamics of S is a GBM with drift r and the no-arbitrage prices (see 6.3) of a call and a put with strike K and maturity T are given by

$$C(t,s) = s\,\Phi(d_1) - Ke^{-r(T-t)}\Phi(d_2), \tag{6.11}$$

$$P(t,s) = Ke^{-r(T-t)}\Phi(-d_2) - s\,\Phi(-d_1), \tag{6.12}$$

where Φ is the standard normal c.d.f. and

$$d_1 = \frac{\log(\frac{s}{K}) + (r + \frac{\sigma^2}{2})(T-t)}{\sigma\sqrt{T-t}}, \qquad d_2 = d_1 - \sigma\sqrt{T-t}.$$

In order to numerically compute the non-trivial liquidation strategy, we solve the variational inequality (VI) of the form

$$\min\left\{ -L_t^\alpha - \mu(t,s)s\,L_s^\alpha - \frac{\sigma^2(t,s)s^2}{2}L_{ss}^\alpha + rL^\alpha - G^\alpha,\ L^\alpha \right\} = 0,$$

$$\tag{6.13}$$

with terminal condition $L^\alpha(T,s) = 0$ and where $(t,s) \in [0,T) \times \mathbb{R}^+$. In Section 6.5, we show that the above VI admits a unique strong solution in the terminology of Bensoussan and Lions (1978) under conditions that include the GBM and exponential OU cases (see Theorem 6.13). For implementation, we adopt the Crank-Nicholson scheme for the VI (7.19) on a finite (discretized) grid $D = [s_{\min}, s_{\max}] \times [0,T]$. We refer to the book by Glowinski (1984) for details on numerical methods for solving inhomogeneous VIs of parabolic type.

6.2.1   *Optimal Liquidation with a GBM Underlying*

We begin our first series of illustrative examples under the GBM model. In view of Proposition 6.2, we observe that, if $\mu \leq r$, it is never optimal to hold a stock or a call (or in general any positive delta position) regardless whether we introduce a risk penalty or not. On the other hand, Proposition 6.2 also implies that, if $\mu > r$ and $\alpha = 0$, it is always optimal to delay.

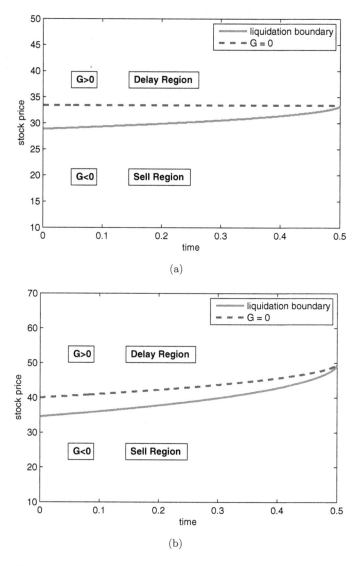

Fig. 6.2 The optimal liquidation boundaries (solid) and the zero contours of G^α (dashed) of a stock (panel (a)) and a call option (panel (b)). We take $T = 0.5$, $r = 0.03$, $\mu = 0.08$, $\sigma = 0.3$, $K = 50$, $\alpha = 0.1$. The loss function is given by $\psi(\ell) = \ell$, with the benchmark $m = 50$ for the stock and $m = C(0, K)$ for the call.

However, with a non-zero risk penalty ($\alpha > 0$), the solution can be non-trivial. To see this, we note that the drive function associated with a call is given by $G^\alpha_{Call}(t, s) = (\mu - r)sC_s - \alpha\psi((m - C(t, s))^+)$ where $C(t, s)$ is the call price in (6.11). In particular, the penalty term is strictly positive at $s = 0$ and decreasing, and it vanishes for large s. On the other hand, the first term $(\mu - r)sC_s$ is strictly increasing from zero at $s = 0$. This implies that there exists a price level $\hat{s}$ such that $G^\alpha_{Call}(t, s)$ is positive in $[0, T] \times [\hat{s}, \infty)$. In turn, it follows from (6.9) that the sell region must be bounded (possibly empty) and the delay region is unbounded. The same argument applies to the case with a stock. Figure 6.2 illustrates this.

Next, we consider the liquidation of a put option. Recall the put price $P(t, s)$ given in (6.12). Its negative Delta implies that for $\mu \geq r$ the drive function $G^\alpha_{Put}(t, s) \leq 0$, $\forall(t, s)$, meaning that it is optimal to sell immediately by Proposition 6.2. In contrast, when $\mu < r$, the sell region is empty if $\alpha = 0$, but under risk penalization the optimal strategy may be non-trivial.

Proposition 6.5. *Consider the optimal liquidation of a put under the GBM model with $\mu < r$ and $\alpha > 0$. Then, the delay region is bounded. Furthermore, it is non-empty if $m < K$ and $\exists \hat{t} \in [0, T]$ such that $\alpha\psi'((m - P(\hat{t}, 0))^+) < r - \mu$.*

Proof. The drive function for the put $G^\alpha \equiv G^\alpha_{Put}(t, s) = (r - \mu)s\Phi(-d_1) - \alpha\psi((m - P(t, s))^+)$ satisfies

$$\lim_{s \to 0} G^\alpha(t, s) = -\alpha\psi((m - K)^+) \leq 0, \tag{6.14}$$

$$\lim_{s \to \infty} G^\alpha(t, s) = -\alpha\psi(m) < 0, \tag{6.15}$$

$$\frac{\partial G^\alpha}{\partial s}(t, s) = [r - \mu - \alpha\psi'((m - P(t, s))^+)\mathbb{1}_{\{m > P(t,s)\}}]\Phi(-d_1)$$
$$\quad - (r - \mu)s\Gamma, \tag{6.16}$$

where $\Gamma = \frac{\partial^2 P}{\partial s^2} \geq 0$. In turn, we fix any $\hat{b} \in (0, \alpha\psi(m))$ and define $\overline{\psi}(\ell) := \min\{\psi(\ell), \hat{b}\}$. This implies the inequality
$$\overline{G}^\alpha(t, s) := (r - \mu)s\Phi(-d_1) - \alpha\overline{\psi}((m - P(t, s))^+) \geq G^\alpha(t, s).$$
Then by Corollary 6.3, we only need to show that $\overline{G}^\alpha$ satisfies the assumptions of Theorem 6.4. We observe that $\overline{G}^\alpha$ is bounded above and it follows from (6.15) that $\lim_{s \to \infty} \overline{G}^\alpha(t, s) \to -\alpha\hat{b} < 0$ for every $t \in [0, T]$. Moreover, there exists $\hat{s} > 0$ such that for every $s > \hat{s}$, $\overline{\psi}((m - P(t, s))^+) = \hat{b}$. Consequently, we have

$$\frac{\partial \overline{G}^\alpha}{\partial t} = (\mu - r)s\phi(d_1)\frac{\log(\frac{s}{K}) - (r + \frac{\sigma^2}{2})(T - t)}{2\sigma(T - t)^{\frac{3}{2}}} \leq 0,$$

for $s > \max\{\hat{s}, K\exp((r+\sigma^2/2)T)\}$ and $t \in [0,T]$. Since $\overline{G}^\alpha(0,s) \to -\alpha\hat{b}$ as $s \to -\infty$, we can choose $b \subset (0, \alpha\hat{b})$ such that $\exists k > \max\{\hat{s}, K\exp((r+\sigma^2/2)T)\}$ and $-b > \overline{G}(0,s) > \overline{G}(t,s)$ in $[0,T] \times [k,\infty)$. Therefore, $\overline{G}$ satisfies the assumptions of Theorem 6.4.

Finally, suppose $\exists \hat{t} \in [0,T]$ such that $\alpha\psi'((m-P(\hat{t},0))^+) < r - \mu$, where $m < K$. It follows from (6.14) and (6.16) that $G^\alpha(\hat{t},0) = 0$ and $\frac{\partial G^\alpha}{\partial s}(\hat{t},0) > 0$, so that the set $\{G^\alpha > 0\}$ is non-empty. In turn, the inclusion (6.9) implies that the delay region is also non-empty.   □

Remark 6.6. As an example, the delay region is empty if

$$\alpha\psi'((m-P(t,0))^+) \geq r - \mu > 0, \qquad \forall t \in [0,T].$$

Indeed, since we have $G^\alpha(t,0) \leq 0$ and $\frac{\partial G^\alpha}{\partial s}(t,0) \leq 0, \forall t \in [0,T]$, G^α cannot be strictly positive. By Proposition 6.2, it is optimal to sell immediately.

Proposition 6.5 is illustrated in Figure 6.3. In these examples, the delay region is non-empty and the sell region is unbounded but may be disconnected (Figure 6.3 (right)). This can arise when, for example, $G^\alpha(t,0) < 0$ for every $t \in [0,T]$, but $\min_t \max G^\alpha(t,s) > 0$. The intuition for a disconnected sell region is as follows. If the put is deeply in the money (i.e. when S_t is close to zero), its market price has very limited room to increase since it is bounded above $Ke^{-r(T-t)}$. At the same time, delaying sale further will incur a penalty. Therefore, when the penalization coefficient α is high, it is optimal to sell at a low stock price level. On the other hand, if the put is deep out of the money (i.e. when S_t is very high), the market price and the Delta of the put are close to zero, meaning the drive function becomes more negative and selling immediately is optimal.

For a long position in calls and the underlying stock, or in puts, the Delta C_s takes a constant sign. As an example of a derivative with a Delta of non-constant sign, we consider a long straddle. This is a combination of a call and a put with strike prices $K_1 \leq K_2$ respectively and the same maturity T. The payoff of a straddle is given by $h^{STD}(S_T) := (S_T - K_1)^+ + (K_2 - S_T)^+$. The market price of a long straddle, denoted by C^{STD}, is simply the sum of the respective Black-Scholes call and put prices, i.e. $C^{STD}(t,s) = C(t,s) + P(t,s)$. For simplicity, we set $K_1 = K_2 = K$.

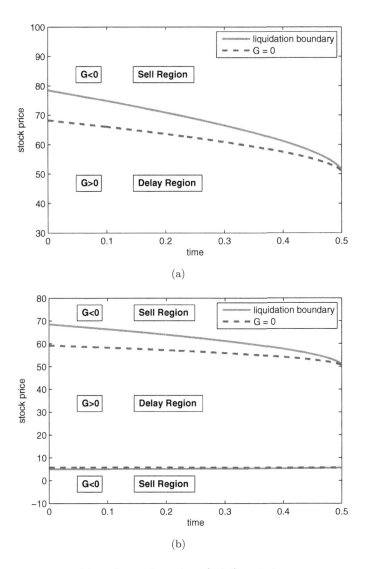

Fig. 6.3 The optimal liquidation boundary (solid) and the zero contour of G^α (dashed) of a put option under GBM dynamics with the loss function $\psi(\ell) = \ell$. We take $m = 2K$, $\alpha = 0.001$ in panel (a), and $m = P(0, K)$, $\alpha = 0.01$ in panel (b). Parameters: $T = 0.5$, $r = 0.03$, $\mu = 0.02$, $\sigma = 0.3$, $K = 50$.

Proposition 6.7. *For the optimal liquidation of a long straddle position under the GBM model, it follows that*
(i) if $\mu = r$, the delay region must be empty;
(ii) if $\mu > r$, the delay region is unbounded;
(iii) if $\mu < r$, the delay region is bounded.

Proof. The straddle's drive function is $G_{STD}^{\alpha}(t,s) = (\mu - r)sC_{s}^{STD}(t,s) - \alpha\psi((m - C^{STD}(t,s))^{+})$. For $\mu = r$, the conclusion follows immediately by Proposition 6.2. If $\mu > r$ we simply notice that $G_{STD}^{\alpha}(t,s) \to \infty$ as $s \to \infty$ for every $t \in [0,T]$, and the assertion follows from the inclusion (6.9).

Now suppose $\mu < r$. We will show that G^{α} satisfies the assumptions of Theorem 6.4. Clearly, G_{STD}^{α} is bounded above. Since $C^{STD}(t,s) \to \infty$ as $s \to \infty$ for every $t \in [0,T]$, then there exists $\hat{s} > 0$ such that, for every $s > \hat{s}$ and $t \in [0,T]$, $\psi((m - C^{STD}(t,s))^{+}) = 0$. Moreover, for $s > \max\{\hat{s}, K\exp\left((r + \sigma^{2}/2)T\right)\}$, we have

$$\frac{\partial\Phi(d_1)}{\partial t} = \phi(d_1)\frac{\log(\frac{s}{K}) - (r + \frac{\sigma^{2}}{2})(T - t)}{2\sigma(T - t)^{\frac{3}{2}}} > 0,$$

and thus

$$\frac{\partial G_{STD}^{\alpha}}{\partial t}(t,s) = 2(\mu - r)s\frac{\partial\Phi(d_1)}{\partial t} \leq 0.$$

This implies $G_{STD}^{\alpha}(0,s) \geq G_{STD}^{\alpha}(t,s)$ for every $t \in [0,T]$. Since $G_{STD}^{\alpha}(0,s) \to -\infty$ as $s \to \infty$, for a fixed $b > 0$ there exists $k_b > 0$ such that $G_{STD}^{\alpha}(0,s) < -b$ for every $s \geq k_b$. Therefore, setting $k = \max\{\hat{s}, K\exp\left((r + \sigma^{2}/2)T\right), k_b\}$, we have $G_{STD}^{\alpha}(t,s) \leq G_{STD}^{\alpha}(0,s) < -b$ in $[0,T] \times [k,\infty)$. Therefore, the assumptions of Theorem 6.4 are satisfied and we conclude. $\qquad\square$

In particular, Proposition 6.7 suggests that when $\mu < r$, the sell region is unbounded, even if $\alpha = 0$. In Figure 6.4, we illustrate the optimal liquidation boundaries for cases (ii) and (iii). When the investor is bullish (panel (a): $\mu = 0.08 > 0.03 = r$), the liquidation boundary is increasing and the delay region is on top of the sell region. Interestingly, the opposite is observed when the investor is bearish (panel (b): $\mu = 0.02 < 0.03 = r$).

We end this section by discussing the liquidation timing of a stock with an infinite horizon ($T = \infty$). This leads to the following stationary optimal stopping problem

$$L(s) = \sup_{\tau \in \mathcal{T}} \mathbb{E}_{s}\left\{\int_{0}^{\tau} e^{-ru}G^{\alpha}(S_u)\,du\right\}. \tag{6.17}$$

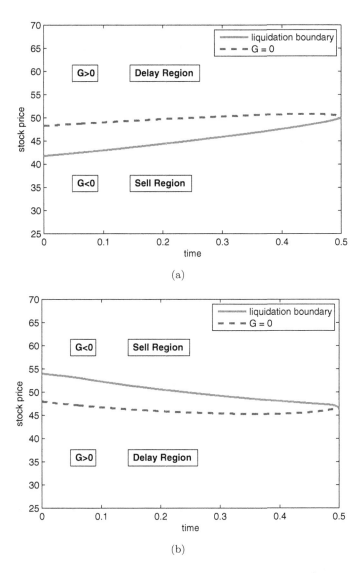

Fig. 6.4 Optimal liquidation boundary and the zero contour of G^α for a straddle under the GBM model with the loss function $\psi(\ell) = \ell$. We set $K = 50$, $m = C^{STD}(0, K)$, $\alpha = 0.1$, $r = 0.03$, $\mu = 0.08$ (panel (a)) and $\mu = 0.02$ (panel (b)).

where $G^\alpha(s) = (\mu - r)s - \alpha\psi((m - s)^+)$ and $\mathcal{T}$ is the set of $\mathbb{F}$-stopping times taking values in $[0, \infty]$. When $\mu \le r$, selling immediately is optimal according to Proposition 6.2, as for the case with finite maturity. As it turns out, the liquidation problem has the opposite trivial solution when $\mu > r$, that is, it is optimal to hold forever.

Proposition 6.8. *If $\mu > r$, then the value function $L(s)$ in (6.17) is infinite and it is optimal to never sell the stock.*

Proof. Consider a candidate stopping time $\tau = \infty$. Then, by applying Tonelli's theorem, we have

$$
\mathbb{E}_s \left\{ \int_0^\infty e^{-ru} G^\alpha(S_u) du \right\}
$$

$$
= \mathbb{E}_s \left\{ \int_0^\infty e^{-ru}(\mu - r)S_u du - \alpha \int_0^\infty e^{-ru}\psi((m - S_u)^+)du \right\}
$$

$$
= \int_0^\infty e^{(\mu-r)u}(\mu - r)s du - \alpha \int_0^\infty e^{-ru}\mathbb{E}\left\{ \psi((m - S_u)^+) \right\} du
$$

$$
\geq \int_0^\infty e^{(\mu-r)u}(\mu - r)s du - \alpha \int_0^\infty e^{-ru}\psi(m)du = \infty,
$$

since $\mu > r$ and ψ is increasing. Hence, $L(s) = \infty$ and it is never optimal to sell. □

6.2.2　*Optimal Liquidation with an Exponential OU Underlying*

In the exponential OU model, the stock price satisfies the SDE

$$
dS_t = \beta(\theta - \log S_t)S_t \, dt + \sigma S_t dW_t,
$$

with $\theta \in \mathbb{R}$ and $\beta, \sigma > 0$. Therefore, the optimal liquidation premium $L(t, s)$ is given by equation (6.7) with the drive function

$$
G^\alpha(t, s) = [\beta(\theta - \log(s)) - r]sV_s(t, s) - \alpha\psi((m - V(t, s))^+), \quad (6.18)
$$

where $V(t, s)$ is a generic option price in (6.3).

In contrast to the GBM case, the optimal liquidation strategy can now be non-trivial for a stock or a call when there is no penalty. More generally, we can prove that the delay region is in fact bounded. The intuition should be clear: when S_t is very high, it is expected to revert back to its long-term mean, so that selling immediately becomes optimal.

Proposition 6.9. *Under the exponential OU model, the delay region for a call is bounded.*

Proof. The drive function G^α_{Call} for the call is given by (6.18) with $V(t, s) = C(t, s)$ (see (6.11) for the call price). It is bounded above, so it satisfies condition (i) of Theorem (6.4). As is well known, the call price satisfies $\frac{\partial C(t,s)}{\partial t} \leq 0$. In addition, $\beta(\theta - \log(s)) - r \leq 0$ iff $s \geq \exp(\theta - \frac{r}{\beta})$, and $\frac{\partial \Phi(d_1)}{\partial t} \geq 0$ for $s \geq K \exp\left((r + \sigma^2/2)T\right)$. In turn, we have

$$\frac{\partial G^\alpha_{Call}}{\partial t}(t, s) = [\beta(\theta - \log(s)) - r]s \frac{\partial \Phi(d_1)}{\partial t}$$
$$+ \alpha \frac{\partial C(t, s)}{\partial t} \psi'((m - C(t, s))^+) \mathbf{1}_{\{m > C(t,s)\}} \leq 0,$$

for $s > \max\{\exp(\theta - \frac{r}{\beta}), K \exp\left((r + \sigma^2/2)T\right)\}$ and $t \in [0, T]$. This implies $G^\alpha_{Call}(0, s) \geq G^\alpha_{Call}(t, s)$. Fix $b > 0$. Since $G^\alpha_{Call}(0, s) \to -\infty$, $\exists k_b > 0$ s.t., $\forall s > k_b$, $G^\alpha_{Call}(0, s) < -b$. Hence, if we set $k = \max\{\exp(\theta - \frac{r}{\beta}), K \exp\left((r + \sigma^2/2)T\right), k_b\}$, we are guaranteed that $G^\alpha_{Call}(t, s) \leq G^\alpha_{Call}(0, s) < -b$ in $[0, T] \times [k, \infty)$, thus satisfying condition (ii) of Theorem 6.4. As a result, Theorem 6.4 applies and gives the boundedness of the delay region for a call. $\square$

Since a stock can be viewed as a call with strike $K = 0$, Proposition 6.6 also applies to the optimal liquidation of a stock over a finite time horizon. Also, we notice the delay region can be empty, and we can identify this case by finding the maximum of the drive function. As an example, we consider the case of the stock with penalty function $\psi((m - S_t)^+) = (m - S_t)^+$, and we obtain the maximizer of G^α in different scenarios

$$\arg\max G^\alpha = \begin{cases} \exp(\theta - 1 - \frac{r-\alpha}{\beta}) & \text{if } \exp(\theta - 1 - \frac{r-\alpha}{\beta}) < m, \\ \exp(\theta - 1 - \frac{r}{\beta}) & \text{if } \exp(\theta - 1 - \frac{r}{\beta}) > m, \\ m & \text{otherwise,} \end{cases}$$

and the corresponding maximum values

$$\max G^\alpha = \begin{cases} (\beta - \alpha)\hat{s}_1 - \alpha(m - s_1^*) & \text{if } \hat{s}_1 < m, \\ \beta\hat{s}_2 & \text{if } \hat{s}_2 > m, \\ (\beta(\theta - \log(m)) - r)m & \text{otherwise,} \end{cases}$$

where

$$\hat{s}_1 = \exp(\theta - 1 - \frac{r - \alpha}{\beta}), \qquad \hat{s}_2 = \exp(\theta - 1 - \frac{r}{\beta}).$$

Thus, the delay region is non-empty if and only if $\max G^\alpha > 0$.

The optimal liquidation boundary for stock is shown in Figure 6.5 for $\alpha = 0$ (panel (a)) and $\alpha > 0$ (panel (b)). We notice that, in both cases, the optimal strategy is to sell immediately if S_t is high enough. Intuitively, if S_t is high, it is expected to revert back to its long-term mean, so selling immediately becomes optimal. However, if S_t is low, the optimal behavior depends on the parameter α. On one hand, S_t is expected to increase and thus the investor should wait to sell at a better price (Figure 6.5(b)). On the other hand, such benefit is countered (if the penalization coefficient is high enough) by the risk incurred from holding the position, and this induces the investor to sell immediately. As a consequence, the sell region is disconnected (Figure 6.5(b)).

Figure 6.6 illustrates the delay region for a call option with penalty. We observe the interesting phenomena where the sell region is connected and contains the nonempty delay region. If the parameter β (which measures the speed of mean reversion) is not sufficiently high, there may be no time for the price of the option to revert back to its long-term mean before expiration, so that selling immediately becomes optimal close to maturity.

Proposition 6.10. *For the liquidation of a put option under the exponential OU model, the delay region is bounded if and only if $\alpha > 0$.*

Proof. The drive function is given by

$$G^\alpha_{Put}(t, s) = [r - \beta(\theta - \log(s))]s\Phi(-d_1) - \alpha\psi((m - P(t, s))^+),$$

If $\alpha = 0$, then we have $\{G^\alpha_{Put} > 0\} = \{s > \exp(\frac{r}{\beta} - \theta)\}$. By (6.9), the delay region contains this set, so it is unbounded.

Now let $\alpha > 0$, and we have the limit

$$\lim_{s \to \infty} G^\alpha_{Put}(t, s) = -\alpha\psi(m) < 0. \qquad (6.19)$$

Next, we fix any $\hat{b} \in (0, \alpha\psi(m))$ and define $\overline{\psi}(\ell) := \min\{\psi(\ell), \hat{b}\}$. With this, we have

$$\overline{G}^\alpha(t, s) := [r - \beta(\theta - \log(s))]s\Phi(-d_1) - \alpha\overline{\psi}\left((m - P(t, s))^+\right) \geq G^\alpha_{Put}(t, s).$$

We observe that $\overline{G}^\alpha$ is bounded above and by (6.19) $\lim_{s \to \infty} \overline{G}^\alpha(t, s) \to -\alpha\hat{b} < 0$ for every $t \in [0, T]$. Moreover, there exists $\hat{s} > 0$ such that for every $s > \hat{s}$, $\overline{\psi}((m - P(t, s))^+) = \hat{b}$. As a result, we have

$$\frac{\partial \overline{G}^\alpha}{\partial t} = (\beta(\theta - \log(s)) - r)s\phi(d_1)\frac{\log(\frac{s}{K} - (r + \frac{\sigma^2}{2})(T - t)}{2\sigma(T - t)^{\frac{3}{2}}} \leq 0,$$

for $s > \max\{\hat{s}, \exp(\frac{r}{\beta} - \theta), K\exp((r + \sigma^2/2)T)\}$ and $t \in [0, T]$. Also, we notice that $\overline{G}^\alpha(0, s) \to -\alpha\hat{b}$ as $s \to -\infty$. This allows us to choose a $b \in (0, \alpha\hat{b})$, then there exists $k > \max\{\hat{s}, K\exp((r + \sigma^2/2)T)\}$ such that

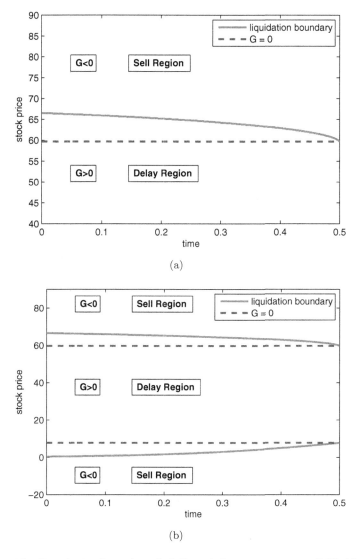

Fig. 6.5   The liquidation boundary (solid) and the zero contour of G^α (dashed) for a stock under exponential OU dynamics. Parameters: $T = 0.5$, $r = 0.03$, $\theta = \log(60)$, $\beta = 4$, $\sigma = 0.3$, $\psi(\ell) = \ell$, $\alpha = 0$ (panel (a)), $\alpha = 1.5$ (panel (b)).

$-b > \overline{G}(0, s) > \overline{G}(t, s)$ in for $(t, s) \in [0, T] \times [k, \infty)$. Therefore, $\overline{G}$ satisfies the assumptions of Theorem 6.4. By Corollary 6.3, we conclude the boundedness of the delay region.   □

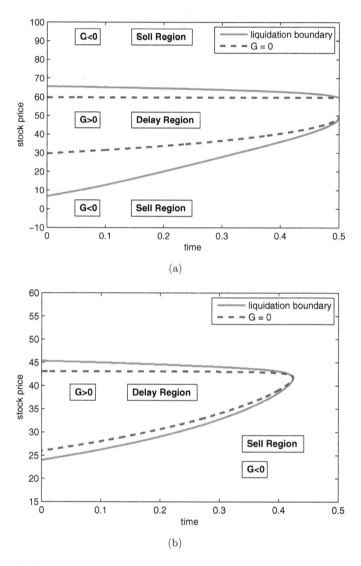

Fig. 6.6   The liquidation boundary (red solid) and the zero contour of G^α (dashed)
for a call under exponential OU dynamics. We take $\alpha = 0.2$, $\theta = \log(60)$, $\beta = 4$
in panel (a), and $\alpha = 0.001$, $\theta = \log(50)$ and $\beta = 0.2$ in panel (b), with common
parameters $T = 0.5$, $r = 0.03$, $\sigma = 0.3$, $\psi(\ell) = \ell$.

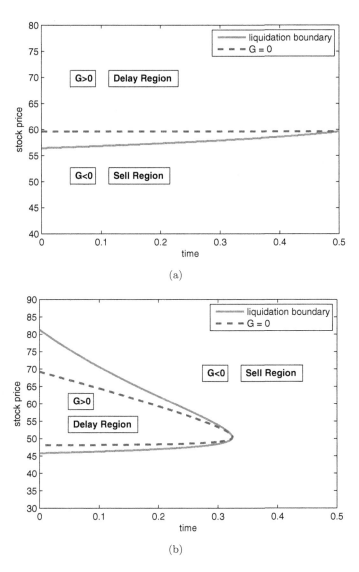

Fig. 6.7 The liquidation boundary (solid) and the zero contour of G^{α} (dashed) for a put option under the exponential OU model. We take $\alpha = 0$ and $K = 50$ in panel (a), and $\alpha = 0.01$ and $K = 40$ in panel (b). Common parameters: $T = 0.5$, $r = 0.03$, $\sigma = 0.3$, $\beta = 4$ and $\theta = \log(60)$, $\psi(\ell) = \ell$.

However, this is no longer true when we incorporate a non-zero risk penalty which reduces the value of waiting. As a result, the holder may sell the put at high and low stock prices. In fact, if the penalization coefficient is large and/or when the time-to-maturity is very short, the optimal liquidation premium may be zero at all stock price levels, resulting in an empty delay region (Figure 6.7(b)).

Proposition 6.10 is illustrated in Figure 6.7. When S_t is low, it is expected to revert back to the (higher) long-term mean, and the put price will decrease. This generates an incentive to sell at a low stock price level. If $\alpha = 0$, when S_t is high, there is no reason to sell since the put price is very low and expected to increase. Consequently, the delay region is on top of the sell region (Figure 6.7(a)).

6.3 Quadratic Penalty

As a variation to the shortfall-based penalty, we consider a risk penalty based on the realized variance of the option price process from the starting time up to the liquidation time. Precisely, the investor now faces the penalized optimal stopping problem

$$
\tilde{J}^\alpha(t,s) := \sup_{\tau \in \mathcal{T}_{t,T}} \mathbb{E}_{t,s}\left\{ e^{-r(\tau-t)} V(\tau, S_\tau) - \alpha \int_t^\tau e^{-r(u-t)} d[V,V]_u \right\}
$$

$$
= \sup_{\tau \in \mathcal{T}_{t,T}} \mathbb{E}_{t,s}\left\{ e^{-r(\tau-t)} V(\tau, S_\tau) \right.
$$

$$
\left. - \alpha \int_t^\tau e^{-r(u-t)} \sigma^2(u, S_u) S_u^2 V_s^2(u, S_u) du \right\},
$$

where $[V,V]$ denotes the quadratic variation of option price process V defined in (6.3). Figure 6.8 illustrates the realized quadratic penalty associated with a simulated call option price path. Compared to the shortfall penalty in Figure 6.1, the realized quadratic penalty is increasing at all times, even when the option price is above its initial price.

Following (6.5), we define the optimal liquidation premium by $\tilde{L}^\alpha(t,s) := \tilde{J}^\alpha(t,s) - V(t,s)$. Again, we shall discuss the stock or option liquidation problems under the GBM and exponential OU models.

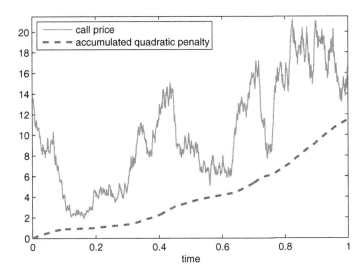

Fig. 6.8 Realized quadratic penalty (dashed) based on a simulated price path (solid) of a call under the GBM model with $\alpha = 0.05$. The price path and other parameters are the same as Figure 6.1.

6.3.1 *Optimal Timing to Sell a Stock*

We first consider the liquidation of a stock with the GBM dynamics in terms of the perpetual optimal stopping problem:

$$\tilde{L}^{\alpha}(s) := \sup_{\tau \in \mathcal{T}} \mathbb{E}_s \left\{ \int_0^{\tau} e^{-ru} \tilde{G}^{\alpha}(S_u) \, du \right\}, \tag{6.20}$$

with the drive function $\tilde{G}^{\alpha}(s) := (\mu - r)s - \alpha \sigma^2 s^2$. If $\mu \leq r$, then selling immediately is always optimal since $\tilde{G}^{\alpha}$ is always negative. In contrast if $\mu > r$, then we obtain a non-trivial closed-form solution.

Theorem 6.11. *Let $\mu > r$. The value function $\tilde{L}^{\alpha}(s)$ in (6.20) is given by the formula*

$$\tilde{L}^{\alpha}(s) = \left\{ \frac{(s^*)^{1-\lambda}}{2 - \lambda} s^{\lambda} - s + B s^2 \right\} \mathbf{1}_{\{s \leq s^*\}}, \tag{6.21}$$

where

$$B = \frac{\alpha \sigma^2}{2\mu + \sigma^2 - r}, \quad \lambda = \frac{1}{\sigma^2} \left[\frac{\sigma^2}{2} - \mu + \sqrt{\left(\frac{\sigma^2}{2} - \mu \right)^2 + 2r\sigma^2} \right], \tag{6.22}$$

$$s^* = \frac{1 - \lambda}{(2 - \lambda)B}, \tag{6.23}$$

and the stopping time $\tau^* = \inf\{t \geq 0 : S_t \geq s^*\}$ *is optimal for (6.20).*

Proof. We first show that (6.21) is the solution of

$$\min\left\{r\Lambda(s) - \mu s\Lambda'(s) - \frac{\sigma^2 s^2}{2}\Lambda''(s) - \tilde{G}^\alpha(s), \Lambda(s)\right\} = 0, \quad s > 0,$$

with $\Lambda(0) = 0$. To do this, we split $\mathbb{R}^+$ into two regions: $\mathcal{D}_1 = (0, s^*)$ and $\mathcal{D}_2 = [s^*, \infty)$ with $s^* > 0$ to be determined. We conjecture that $\Lambda(s) = 0$ in $\mathcal{D}_2$, and for $s \in \mathcal{D}_1$ $\Lambda(s)$ solves

$$r\Lambda(s) - \mu s\Lambda'(s) - \frac{\sigma^2 s^2}{2}\Lambda''(s) - \tilde{G}^\alpha(s) = 0. \tag{6.24}$$

By direct substitution, the general solution to equation (6.24) is of the form

$$\Lambda(s) = C_1 s^{\lambda_1} + C_2 s^{\lambda_2} - s + Bs^2,$$

where C_1 and C_2 are constants to be determined, B is specified in (6.22) and

$$\lambda_k = \frac{1}{\sigma^2}\left[\frac{\sigma^2}{2} - \mu + (-1)^k \sqrt{\left(\mu - \frac{\sigma^2}{2}\right)^2 + 2r\sigma^2}\right], \quad k \in \{1, 2\}.$$

We apply the continuity and smooth pasting conditions at $s = 0$ and $s = s^*$ to get

$$\lim_{s \downarrow 0} \Lambda(s) = 0 \Rightarrow C_1 = 0,$$

$$\lim_{s \uparrow s^*} \Lambda(s) = 0 \Rightarrow C_2(s^*)^{\lambda_2} - s^* + B(s^*)^2 = 0, \tag{6.25}$$

$$\lim_{s \uparrow s^*} \Lambda'(s) = 0 \Rightarrow \lambda_2 C_2(s^*)^{\lambda_2 - 1} - 1 + 2Bs^* = 0. \tag{6.26}$$

Solving the system of equations (6.25)–(6.26) gives C_2 and s^* as in (6.22)–(6.23). One can verify by substitution that $\Lambda(s)$ is indeed a classical solution of (6.24).

By Ito's formula and (6.24), $(\Lambda(S_t))_{t\geq 0}$ is a $(\mathbb{P}, \mathbb{F})$-supermartingale, so for every $\mathbb{F}$-stopping time τ and $n \in \mathbb{N}$, we have

$$\Lambda(s) \geq \mathbb{E}_{0,s}\left\{\int_0^{\tau \wedge n} e^{-ru}\tilde{G}^\alpha(S_u)du\right\}. \tag{6.27}$$

Maximizing (6.27) over τ and n yields that $\Lambda(s) \geq \tilde{L}^\alpha(s)$ for $s \geq 0$. The reverse inequality is deduced from the probabilistic representation $\Lambda(s) = \mathbb{E}_{0,s}\left\{\int_0^{\tau^*} e^{-ru}\tilde{G}^\alpha(S_u)du\right\}$, with the candidate stopping time $\tau^* := \inf\{t \geq 0 : S_t \geq s^*\}$. Hence, we conclude that $\Lambda(s) = \tilde{L}^\alpha(s)$ and τ^* is optimal. $\square$

We see that the optimal liquidation threshold s^* in (6.23) is non-negative if and only if $\lambda_2 < 1$, which is equivalent to the condition $\mu > r$ in Theorem 6.11. Otherwise, $\tilde{L}^\alpha(s) = 0$ and the optimal strategy is to sell immediately.

In Figure 6.9, we illustrate the optimal liquidation premium $\tilde{L}^\alpha(s)$ for various values of μ and σ. As μ increases, the optimal threshold as well as the optimal liquidation premium (at all stock price levels) increase. On the other hand, a higher volatility reduces the optimal liquidation premium at every initial stock price. We also observe that $\tilde{L}^\alpha(s)$ smooth-pastes the level 0 at the optimal threshold s^*, as is expected from (6.25) and (6.26).

If S follows the exponential OU dynamics, the drive function for liquidating a stock is

$$\tilde{G}^\alpha(s) = [\beta(\theta - \log(s)) - r - \alpha s]s. \tag{6.28}$$

In this case, we do not have a closed-form solution. Nevertheless we observe from (6.28) that the delay region is non-empty, namely, $\{\tilde{L}^\alpha > 0\} \supseteq \{s < \tilde{s}\}$, where $\tilde{s}$ is determined uniquely from the equation

$$\beta(\theta - \log(\tilde{s})) - r - \alpha\tilde{s} = 0.$$

On the other hand, since $\tilde{G}^\alpha \to -\infty$ as $s \to \infty$, we expect intuitively that the investor will sell when the stock price is high.

6.3.2 *Liquidation of Options*

We now discuss some numerical examples to demonstrate the liquidation strategies for European call and put options. With strike K and maturity T, the drive functions are respectively given by

$$\tilde{G}^\alpha_{Call}(t, s) = s\Phi(d_1)\big(\mu - r - \alpha\sigma^2 s\Phi(d_1)\big), \tag{6.29}$$

$$\tilde{G}^\alpha_{Put}(t, s) = s\Phi(-d_1)\big(r - \mu - \alpha\sigma^2 s\Phi(-d_1)\big). \tag{6.30}$$

When $\mu \leq r$ and $\alpha > 0$, the drive function $\tilde{G}^\alpha_{Call}(t, s)$ is negative for all (t, s), so it is optimal to sell the call immediately. However, when $\mu > r$ and $\alpha > 0$, we notice from (6.29) that, when the stock price is sufficiently large (resp. small), the drive function of a call is negative (resp. positive). Hence, as we see in Figure 6.10, it is optimal to sell the call when the stock price is high, and the optimal liquidation boundary is lower as the penalization coefficient increases. In contrast to the shortfall penalty, the investor now is subject to a higher penalty when the stock price is high under the quadratic penalty. Consequently, the sell region is now above the delay region, as opposed to being at the bottom in the shortfall case in Figure 6.2(b).

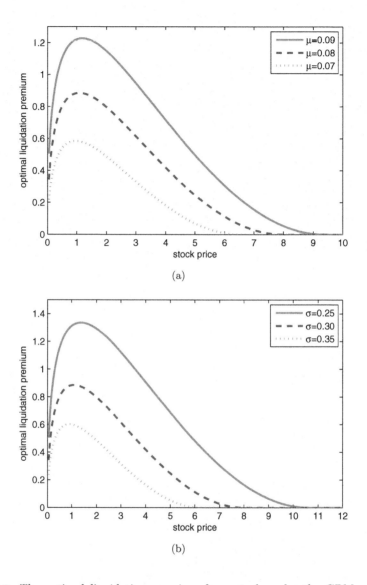

Fig. 6.9 The optimal liquidation premium for a stock under the GBM model for different values of μ and σ. In panel (a), we take $r = 0.03$, $\sigma = 0.3$ and $\alpha = 0.2$, and the liquidation threshold $s^* = 9.37, 7.97, 6.52$ for $\mu = 0.09, 0.08, 0.07$ respectively. In panel (b), we take $r = 0.03$, $\mu = 0.08$, and $\alpha = 0.1$, and the liquidation threshold $s^* = 10.63, 7.97, 6.26$ for $\sigma = 0.25, 0.30, 0.35$.

In the put option case, we observe from (6.30) that

$$\lim_{s \to 0} r - \mu - \alpha \sigma^2 s \Phi(-d_1) = \lim_{s \to \infty} r - \mu - \alpha \sigma^2 s \Phi(-d_1) = r - \mu.$$

Consequently, when $\mu < r$ and the stock price is sufficiently large or small, the drive function is strictly positive and it is optimal to hold the position. In contrast, the shortfall converges to $\psi(m) > 0$ as s increases (see (6.15)), which means that it is optimal to sell when the stock price is high (see Figure 6.3). We illustrate the timing strategies under quadratic penalty in Figure 6.10 (right). As expected there is a low and a high delay regions which are separated by a sell region in the middle. Also we notice that as the penalization coefficient α increases, the sell region expands.

Under the exponential OU model, the drive functions for selling a call and a put are, respectively,

$$\tilde{G}^\alpha_{Call}(t, s) = s \Phi(d_1) \big(\theta - r - \beta \log s - \alpha \sigma^2 s \Phi(d_1) \big),$$

$$\tilde{G}^\alpha_{Put}(t, s) = s \Phi(-d_1) \Big(r - \theta + \beta \log s - \alpha \sigma^2 s \Phi(-d_1) \Big).$$

In Figure 6.11, we can visualize the optimal liquidation premium $\tilde{L}^\alpha(t, s)$ for a call (panel (b)) and a put (panel (a)). In the call case, the delay region, which corresponds to the area where $\tilde{L}^\alpha > 0$, is bounded. When s is sufficiently high, $\tilde{L}^\alpha$ vanishes and it is optimal to sell. This is intuitive since $\lim_{s \to \infty} \tilde{G}^\alpha_{Call}(t, s) = -\infty$ and $\tilde{G}^\alpha_{Call}$ is positive for sufficiently small s.

In contrast, the drive function for the put $\tilde{G}^\alpha_{Put}(t, s)$ is negative when $s < \exp\left(\frac{\theta - r}{\beta}\right)$, for every $\alpha \geq 0$. Therefore, as Figure 6.11 indicates, one expects the optimal liquidation premium to vanish for small s, so the investor will sell when the put price is high. Compared to Figure 6.7 with a shortfall penalty, the investor does not sell when the underlying stock price is very high. This is because the drive function $\tilde{G}^\alpha_{Put}(t, s)$ stays positive for large s (recall (6.9)). As time approaches maturity, the delay liquidation premium decreases to its terminal condition of value zero.

6.4 Concluding Remarks

In summary, we have provided a flexible mathematical model for the optimal liquidation of option positions under a path-dependent penalty. We have identified the situations where the optimal timing is trivial, and solved for non-trivial liquidation strategy via variational inequality. The penalty type as well as the penalization coefficient can give rise to very different liquidation timing. Our findings are useful for both individual and institutional investors who use options for speculative investments or risk management purposes.

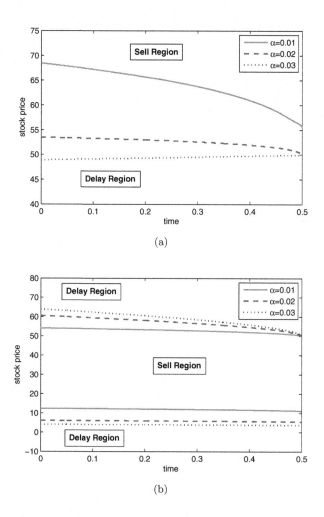

(a)

(b)

Fig. 6.10   The liquidation boundaries for a call option (panel (a)) and a put option (panel (b)) under the GBM mode with different values of α. Parameters: $T = 0.5$, $r = 0.03$, $\sigma = 0.3$, $K = 50$, $\mu = 0.08$ (call) and $\mu = 0.02$ (put).

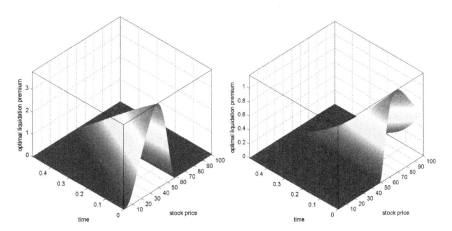

Fig. 6.11 The optimal liquidation premium for a call option (left) and a put option (right) with exponential OU dynamics. We take $T = 0.5$, $r = 0.03$, $\sigma = 0.3$, $K = 50$, $\alpha = 0.1$, $\beta = 4$ and $\theta = \log(60)$.

For future research, a natural direction is to adapt our model to the problem of sequentially buying and selling an option. Moreover, one can consider applying the methodology to derivatives other than equity options. For example, the liquidation of credit derivatives with pricing measure discrepancy but without risk penalty is discussed in Chapter 7. It would be both mathematically interesting and challenging to study option liquidation under incomplete markets. On the other hand, our model can be extended to markets with liquidity cost and price impact (see e.g. Almgren (2003); Lorenz and Almgren (2011); Schied and Schöneborn (2009)). Finally, the path-dependent risk penalization can also be incorporated to dynamic portfolio optimization problems to account for adverse performance during the investment horizon.

6.5 Strong Solution to the Inhomogeneous Variational Inequality

In this section, we follow the terminology and procedures in Bensoussan and Lions (1978), and establish the existence and uniqueness of a strong solution to the variational inequality (7.19) under conditions that are applicable to the GBM and exponential OU models.

6.5.1 *Preliminaries*

We express prices in logarithmic scale by setting $X_t = \log(S_t)$. Equation (6.1) then becomes

$$dX_t = \eta(t, X_t)dt + \kappa(t, X_t)dW_t,$$

for some functions $\kappa(t, x)$ and $\eta(t, x)$. Next, we define the operator $\mathcal{A}$ by

$$
\begin{aligned}
\mathcal{A}[\cdot] &= -\frac{\kappa^2(t, x)}{2}\frac{\partial^2 \cdot}{\partial x^2} - \eta(t, x)\frac{\partial \cdot}{\partial x} + r \cdot \\
&= -\frac{\partial \cdot}{\partial x}\left(a_2(t, x)\frac{\partial \cdot}{\partial x}\right) + a_1(t, x)\frac{\partial \cdot}{\partial x} + r \cdot,
\end{aligned}
$$

where

$$a_1(t, x) = \frac{1}{2}\frac{\partial}{\partial x}\kappa^2(t, x) - \eta(t, x), \qquad a_2(t, x) = \frac{\kappa^2(t, x)}{2}.$$

In term of log-prices, we express the drive function as $g(t, x) = G^\alpha(t, e^x)$ and the optimal liquidation premium as $u(t, x) = L(t, e^x)$. Throughout, we denote the domain $\mathcal{D} = [0, T] \times \mathbb{R}$. In order to solve the VI (7.19), it is equivalent to solve the VI problem:

$$
\begin{cases}
-\frac{\partial u}{\partial t} + \mathcal{A}[u] - g(t, x) \geq 0, \ u(t, x) \geq 0, & (t, x) \in \mathcal{D}, \\
\left(-\frac{\partial u}{\partial t} + \mathcal{A}[u] - g(t, x)\right) u = 0, & (t, x) \in \mathcal{D}, \\
u(T, x) = 0, & x \in \mathbb{R}.
\end{cases} \tag{6.31}
$$

We describe an appropriate class of solutions for (6.31) in a suitable Sobolev space and prove that such a solution exists and is unique. First, let us define, for $\lambda(x) = \exp(-n|x|)$, $n \in \mathbb{N}$,

$$\mathcal{L}_\lambda^2(\mathbb{R}) = \{v \mid \sqrt{\lambda}v \in L^2(\mathbb{R})\},$$

$$\mathcal{H}_\lambda^1(\mathbb{R}) = \{v \in L_\lambda^2(\mathbb{R}) \mid \frac{\partial v}{\partial x} \in L_\lambda^2(\mathbb{R})\},$$

$$\mathcal{H}_{0,\lambda}^1(\mathbb{R}) = \{v \in H_\lambda^1(\mathbb{R}) \mid \lim_{|x| \to \infty} v(x) = 0\}.$$

These are Hilbert spaces when endowed with the following inner products

$$(f, g)_{L^2} = \int_\mathbb{R} \lambda f g\, dx, \quad f, g \in L_\lambda^2(\mathbb{R}),$$

$$(f, g)_{H^1} = \int_\mathbb{R} \lambda f g\, dx + \int_\mathbb{R} \lambda \frac{\partial f}{\partial x}\frac{\partial g}{\partial x}\, dx, \quad f, g \in H_\lambda^1(\mathbb{R}).$$

We denote by $\mathcal{H}^1_{c,\lambda}(\mathbb{R})$ the set of functions $w \in \mathcal{H}^1_\lambda(\mathbb{R})$ with compact support. For $u \in \mathcal{H}^1_{0,\lambda}(\mathbb{R})$, $w \in \mathcal{H}^1_{c,\lambda}(\mathbb{R})$, we define the operator

$$\mathcal{I}_\lambda(t, u, w) = \int_{\mathbb{R}} a_2(t, x) \left(\lambda \frac{\partial u}{\partial x} \frac{\partial w}{\partial x} + w \frac{\partial u}{\partial x} \frac{\partial \lambda}{\partial x} \right) dx$$
$$+ \int_{\mathbb{R}} a_1(t, x) \lambda \frac{\partial u}{\partial x} w \, dx + r \int_{\mathbb{R}} \lambda u w \, dx.$$

We can assume without loss of generality (see Sect. 3.2.17 Bensoussan and Lions (1978)) that $\mathcal{I}_\lambda$ is coercive on $\mathcal{H}^1_{c,\lambda}(\mathbb{R})$, i.e.

$$\mathcal{I}_\lambda(t, w, w) \geq \alpha \|w\|_{H^1} \quad \forall w \in \mathcal{H}^1_{c,\lambda}(\mathbb{R}), \, \alpha > 0.$$

Integrating by parts allows us to extend $\mathcal{I}_\lambda$ to a bilinear form on the whole space $\mathcal{H}^1_{0,\lambda}(\mathbb{R})$. In particular, we set

$$\mathcal{I}_\lambda(t, u, v) = \int_{\mathbb{R}} \left[a_2(t, x) \lambda \frac{\partial u}{\partial x} \frac{\partial v}{\partial x} + a_2(t, x) \frac{\partial \lambda}{\partial x} \frac{\partial u}{\partial x} v \right] dx$$
$$+ \int_{\mathbb{R}} \left(r - \frac{1}{2} \frac{\partial a_1}{\partial x} - \frac{1}{2\lambda} a_1 \frac{\partial \lambda}{\partial x} \right) \lambda u v \, dx.$$

with $u, v \in \mathcal{H}^1_{0,\lambda}(\mathbb{R})$.

Following Section 5.9.2 of Evans (1998) and Section 2.6 of Bensoussan and Lions (1978), we define the space $\mathcal{L}^p(0, T; X)$ consisting of all strongly measurable functions $\chi : [0, T] \to X$ with

$$\|\chi\|_{\mathcal{L}^p(0,T;X)} = \left(\int_0^T \|\chi(t)\|_X^p \, dt \right)^{1/p}, \quad 1 \leq p < \infty,$$

and for $p = \infty$,

$$\|\chi\|_{\mathcal{L}^\infty(0,T;X)} = \operatorname*{ess\,sup}_{0 \leq t \leq T} \|\chi(t)\|_X,$$

For $\chi \in \mathcal{L}^1(0, T; X)$, we say $\nu \in \mathcal{L}^1(0, T; X)$ is the weak derivative of χ, denoted by $\nu = \frac{\partial \chi}{\partial t}$, if

$$\int_0^T \frac{\partial w}{\partial t} \chi(t) dt = - \int_0^T w(t) \nu(t) dt, \quad \forall w \in C_c^\infty([0, T]).$$

The Sobolev space $\mathcal{H}^1(0, T; X)$ consists of all functions $\chi \in \mathcal{L}^2(0, T; X)$ such that the weak derivative exists and belongs to $\mathcal{L}^2(0, T; X)$. Furthermore, we set

$$\|\chi\|_{\mathcal{H}^1(0,T;X)} = \left(\int_0^T \|\chi(t)\|_X^2 + \|\frac{\partial}{\partial t} \chi(t)\|_X^2 \, dt \right)^{1/2}, \tag{6.32}$$

which makes $\mathcal{H}^1(0, T; X)$ an Hilbert space (see Section 5.9.2 in Evans (1998)).

6.5.2   Main Results

Definition 6.12. A function $u : \mathcal{D} \to \mathbb{R}$ is a *strong solution* of problem (6.31) if, $\forall\, v \in \mathcal{H}^1_\lambda(\mathbb{R})$, $v \geq 0$ a.e., the following conditions are satisfied:

$$
\begin{cases}
u \in \mathcal{L}^2(0,T;\mathcal{H}^1_{0,\lambda}(\mathbb{R})),\ \frac{\partial u}{\partial t} \in \mathcal{L}^2(0,T;\mathcal{L}^2_\lambda(\mathbb{R})), \\[4pt]
-\left(\frac{\partial u}{\partial t}, v - u\right) - \mathcal{I}_\lambda(t; u, v - u) \leq (g, v - u), \\[4pt]
u \geq 0 \text{ a.e. in } \mathcal{D}, \\[4pt]
u(T,x) = 0,\ x \in \mathbb{R}.
\end{cases}
\tag{6.33}
$$

We shall impose the following conditions on a_2, a_1, g.

Assumption A. $a_2,\ \frac{\partial a_2}{\partial t}$ and $\frac{\partial a_1}{\partial x} \in \mathcal{L}^\infty(\mathcal{D})$; a_1 and $\frac{\partial a_1}{\partial t} \in C^0(\overline{\mathcal{D}})$; $g \in \mathcal{H}^1(0,T;\mathcal{L}^2_\lambda(\mathbb{R}))$.

Theorem 6.13. *Under Assumption A, the variational inequality in* (6.33) *has a unique strong solution.*

Proof. Assumption A is equivalent to assumptions (2.223), (2.224), (2.238), (2.239), (2.240) of Bensoussan and Lions (1978), and we also follow their Remark 2.24 to use $\lambda(x) = e^{-n|x|}$ for some arbitrarily fixed $n > 0$ in our definition of Hilbert spaces. In turn, we can apply their Theorem 2.21 and our statement follows. $\qquad\square$

Our main objective is to verify that Assumption A is satisfied for our applications so that Theorem 6.13 applies to ensure the existence of a unique strong solution to the VI (7.19). To see this, we first write down the operators associated with the log-price $X_t = \log(S_t)$ under the GBM and exponential OU models, namely,

$$
\mathcal{A}[v] = \frac{\sigma^2}{2}\frac{\partial^2 v}{\partial x^2} + \mu\frac{\partial v}{\partial x}, \qquad \mathcal{A}[v] = \frac{\sigma^2}{2}\frac{\partial^2 v}{\partial x^2} + (\hat{\theta} - \beta x)\frac{\partial v}{\partial x}.
$$

Therefore, $a_2 = \sigma^2/2$ is constant and a_1 is an affine function in x for both cases, so these coefficients meet the requirements in Assumption A.

It remains to verify that the drive function $g(t,x) = G(t,e^x) \in \mathcal{H}^1(0,T;\mathcal{L}^2_\lambda(\mathbb{R}))$. In view of (6.32), we want to show that there exists $n > 0$ such that

$$
\int_0^T \|g(t,x)\|^2_{\mathcal{L}^2_\lambda(\mathbb{R})}\,dt = \int_0^T \int_{\mathbb{R}} \left(g(t,x)e^{-\frac{n}{2}|x|}\right)^2 dx\,dt, \quad \text{and}
$$

$$
\int_0^T \|\frac{\partial g}{\partial t}(t,x)\|^2_{\mathcal{L}^2_\lambda(\mathbb{R})}\,dt = \int_0^T \int_{\mathbb{R}} \left(\frac{\partial g}{\partial t}(t,x)e^{-\frac{n}{2}|x|}\right)^2 dx\,dt
$$

are finite, where

$$g(t, x) = (r - \mu(t, e^x))e^x V_s(t, e^x) - \alpha\psi((m - V(t, e^x))^+),$$

$$\frac{\partial g}{\partial t}(t, x) = (r - \mu(t, e^x))e^x V_{ts}(t, e^x)$$
$$+ \alpha\psi'((m - V(t, e^x))^+)V_t(t, e^x)\mathbf{1}_{\{m > V(t, e^x)\}}.$$

Here, the subscripts of V indicate the partial derivatives in t and s. Recall the drift functions $\mu(t, e^x) = \mu$ under GBM and $\mu(t, e^x) = \beta(\theta - x)$ under exponential OU models. We notice that, in both cases, the drift does not depend on t, so we just write $\mu(e^x)$. Also, we observe that ψ and ψ' are increasing, and $\psi'(\ell)$ is bounded for any finite ℓ. For both call and put options, there exist positive constants h_1, q_1, h_2, q_2 such that $|V_s(t, e^x)| \leq 1, |V_t(t, e^x)| \leq h_1 e^x + q_1, |V_{st}(t, e^x)| \leq h_2 e^x + q_2$. Together, these imply the time-independent bounds for both models:

$$|g(t, x)| \leq |r - \mu(e^x)|e^x + \alpha\psi(m) = o(e^{2|x|}),$$

$$|\frac{\partial g}{\partial t}(t, x)| \leq |r - \mu(e^x)|(h_1 e^x + q_1)e^x + \alpha\psi'(m)(h_2 e^x + q_2) = o(e^{2|x|}).$$

This implies that by choosing $n > 4$, we have

$$\int_0^T \|g(t, x)\|_{\mathcal{L}_\lambda^2(\mathbb{R})}^2 dt \leq \int_0^T \left\| |r - \mu(e^x)|e^x + \alpha\psi(m) \right\|_{\mathcal{L}_\lambda^2(\mathbb{R})}^2 dt < \infty,$$

$$\int_0^T \|\frac{\partial g}{\partial t}(t, x)\|_{\mathcal{L}_\lambda^2(\mathbb{R})}^2 dt$$
$$\leq \int_0^T \left\| |r - \mu(e^x)|(h_1 e^x + q_1)e^x + \alpha\psi'(m)(h_2 e^x + q_2) \right\|_{\mathcal{L}_\lambda^2(\mathbb{R})}^2 dt < \infty.$$

Hence, we conclude that $g \in \mathcal{H}^1(0, T; \mathcal{L}_\lambda^2(\mathbb{R}))$ for both puts and calls under the GBM and exponential OU models, and Assumption A is satisfied.

As a final remark, Sect. 3.4 of Bensoussan and Lions (1978) also provides the probabilistic representation of the strong solution $u(t, x)$ of the VI (6.31), given by

$$u(t, x) = \sup_{\tau \in \mathcal{T}_{t,T}} \mathbb{E}_{t,x} \left\{ \int_t^\tau e^{-r(u-t)} g(u, X_u) \, du \right\}, \tag{6.34}$$

where $dX_u = \eta(u, X_u)du + \kappa(u, X_u)dW_u$ and $X_t = x$. By the definition $L(t, e^x) = u(t, x)$, the optimal stopping problem in (6.34) resembles that for the optimal liquidation premium in (6.7).

Chapter 7

Trading Credit Derivatives

In credit derivatives trading, one important question is how the market compensates investors for bearing credit risk. A number of related studies have examined analytically and empirically the structure of default risk premia inferred from the market prices of corporate bonds, credit default swaps, and multi-name credit derivatives.[1] A major risk premium component is the *mark-to-market risk premium* which accounts for the fluctuations in default risk. In addition, there is the *event risk premium* (or jump-to-default risk premium) that compensates for the uncertain timing of the default event.

When it is not possible to perfectly hedge away all risks, the market is incomplete. In such a market, there exist many different risk-neutral measures that yield no-arbitrage prices. From standard no-arbitrage pricing theory, risk premia specification is inherently tied to the selection of risk-neutral pricing measures. A typical buy-side investor (e.g. hedge fund manager or proprietary trader) would identify trading opportunities by looking for mispriced contracts in the market. This implies selecting a pricing measure to reflect her view on credit risk evolution and the required risk premia. Hence, the investor's pricing measure may differ from that represented by the market prices.

Price discrepancy is also important for investors with credit-sensitive positions who may need to control risk exposure through liquidation. The central issue lies in the timing of liquidation as investors have the option to sell at the current market price or wait for a later opportunity. The optimal strategy, as we will study, depends on the sources of risks, risk premia, as well as derivative payoffs.

---

[1]See Azizpour *et al.* (2011); Berndt *et al.* (2005); Driessen (2005); Jarrow *et al.* (2005) and references therein.

This chapter proposes a new approach to tackle the optimal liquidation problem on two fronts. First, we provide a mathematical framework for price discrepancy between the market and investors under an intensity-based credit risk model. Second, we analyze the optimal stopping problem corresponding to the liquidation of credit derivatives under price discrepancy.

In order to measure the benefit of optimally timing to sell as opposed to immediate liquidation, we employ the concept of *delayed liquidation premium*. It turns out to be a very useful tool for analyzing the optimal stopping problem. The intuition is that the investor should wait as long as the delayed liquidation premium is strictly positive. Moreover, the delayed liquidation premium reveals the roles of risk premia in the liquidation timing. Under a Markovian credit risk model, the optimal timing is characterized by a liquidation boundary solved from a variational inequality. For numerical illustration, we provide examples where the default intensity and interest rate are mean-reverting.

The rest of the chapter is organized as follows. In Section 7.1, we present the mathematical model for price discrepancy and formulate the optimal liquidation problem under a general intensity-based credit risk model. In Section 7.2, we study the problem within a Markovian market and characterize the optimal liquidation strategy for a general defaultable claim. In Section 7.3, we investigate the impact of pricing measure on the investor's liquidation timing for single-name credit derivatives, e.g. defaultable bonds and credit default swaps (CDS). In Section 7.4, we discuss the optimal liquidation of credit default index swap. In Section 7.5, we examine the optimal buy-and-sell strategy for defaultable claims.

7.1   Problem Formulation

This section provides the mathematical formulation of price discrepancy and the optimal liquidation of credit derivatives under an intensity-based credit risk model. We fix a probability space $(\Omega, \mathcal{G}, \mathbb{P})$, where $\mathbb{P}$ is the historical measure, and denote T as the maturity of derivatives in question. There is a stochastic risk-free interest rate process $(r_t)_{0 \leq t \leq T}$. The default arrival is described by the first jump of a doubly-stochastic Poisson process. Precisely, assuming a default intensity process $(\hat{\lambda}_t)_{0 \leq t \leq T}$, we define the default time τ_d by

$$\tau_d = \inf\{t \geq 0 : \int_0^t \hat{\lambda}_s ds > E\}, \qquad \text{where } E \sim Exp\,(1) \text{ and } E \perp \hat{\lambda}, r.$$

The associated default counting process is $N_t = \mathbf{1}_{\{t \geq \tau_d\}}$. The filtration $\mathbb{F} = (\mathcal{F}_t)_{0 \leq t \leq T}$ is generated by r and $\hat{\lambda}$. The full filtration $\mathbb{G} = (\mathcal{G}_t)_{0 \leq t \leq T}$ is defined by $\mathcal{G}_t = \mathcal{F}_t \vee \mathcal{F}_t^N$ where $(\mathcal{F}_t^N)_{0 \leq t \leq T}$ is generated by N.

7.1.1 Price Discrepancy

By standard no-arbitrage pricing theory, the market price of a defaultable claim, denoted by $(P_t)_{0 \leq t \leq T}$, is computed from a conditional expectation of discounted payoff under the market risk-neutral (or equivalent martingale) pricing measure $\mathbb{Q} \sim \mathbb{P}$. In many parametric credit risk models, the market pricing measure $\mathbb{Q}$ is related to the historical measure $\mathbb{P}$ via the default risk premia (see Section 7.2.1 below). Throughout we assume a single historical measure $\mathbb{P}$ shared by all investors, but they differ in their views on the risk premia for various sources of risks. Also, we adopt the standard hypothesis (H) that every $\mathbb{F}$-local martingale is a $\mathbb{G}$-local martingale holds under $\mathbb{Q}$ (see Chapter 8 of Bielecki and Rutkowski (2002)).

We can describe a general defaultable claim by the quadruple (Y, A, R, τ_d), where $Y \in \mathcal{F}_T$ is the terminal payoff if the defaultable claim survives at T, $(A_t)_{0 \leq t \leq T}$ is an $\mathbb{F}$-adapted continuous process of finite variation with $A_0 = 0$ representing the promised dividends until maturity or default, and $(R_t)_{0 \leq t \leq T}$ is an $\mathbb{F}$-predictable process representing the recovery payoff paid at default. Similar notations are used in Bielecki *et al.* (2008) where the following integrability conditions are assumed:

$$\mathbb{E}^{\mathbb{Q}}\{|e^{-\int_0^T r_v dv} Y|\} < \infty, \quad \mathbb{E}^{\mathbb{Q}}\{|\int_{(0,T]} e^{-\int_0^u r_v dv}(1 - N_u)dA_u|\} < \infty, \quad \text{and}$$

$$\mathbb{E}^{\mathbb{Q}}\{|e^{-\int_0^{\tau_d \wedge T} r_v dv} R_{\tau_d \wedge T}|\} < \infty.$$

For a defaultable claim (Y, A, R, τ_d), the associated cash flow process $(D_t)_{0 \leq t \leq T}$ is defined by

$$D_t := Y\mathbf{1}_{\{\tau_d > T\}}\mathbf{1}_{\{t \geq T\}} + \int_{(0, t \wedge T]} (1 - N_u)dA_u + \int_{(0, t \wedge T]} R_u dN_u. \quad (7.1)$$

Then, the (cumulative) market price process $(P_t)_{0 \leq t \leq T}$ is given by the conditional expectation under the market pricing measure $\mathbb{Q}$:

$$P_t := \mathbb{E}^{\mathbb{Q}}\{\int_{(0,T]} e^{-\int_t^u r_v dv} dD_u | \mathcal{G}_t\}.$$

One simple example is the zero-coupon zero-recovery defaultable bond $(1, 0, 0, \tau_d)$, whose market price is simply $P_t = \mathbb{E}^{\mathbb{Q}}\{e^{-\int_t^T r_v dv} \mathbf{1}_{\{\tau_d > T\}} | \mathcal{G}_t\}$.

When a perfect replication is unavailable, the market is incomplete and there exist different risk-neutral pricing measures that give different no-arbitrage prices for the same defaultable claim. Mathematically, this amounts to assigning a different risk-neutral pricing measure $\tilde{\mathbb{Q}} \sim \mathbb{Q} \sim \mathbb{P}$. The investor's reference price process $(\tilde{P}_t)_{0 \leq t \leq T}$ is given by the conditional expectation under investor's risk-neutral pricing measure $\tilde{\mathbb{Q}}$:

$$\tilde{P}_t := \mathbb{E}^{\tilde{\mathbb{Q}}}\{\int_{(0,T]} e^{-\int_t^u r_v dv} dD_u | \mathcal{G}_t\},$$

whose discounted price process $(e^{-\int_0^t r_v dv} \tilde{P}_t)_{0 \leq t \leq T}$ is a $(\tilde{\mathbb{Q}}, \mathbb{G})$-martingale. We assume that the standard hypothesis (H) also holds under $\tilde{\mathbb{Q}}$.

7.1.2   *Delayed Liquidation Premium*

A defaultable claim holder can sell her position at the prevailing market price. If she completely agrees with the market price, then she will be indifferent to sell at any time. Under price discrepancy, however, there is a timing option embedded in the optimal liquidation problem. Precisely, in order to maximize the expected spread between the investor's price and the market price, the holder solves the optimal stopping problem:

$$J_t := \operatorname*{ess\,sup}_{\tau \in \mathcal{T}_{t,T}} \mathbb{E}^{\tilde{\mathbb{Q}}}\{e^{-\int_t^\tau r_v dv}(P_\tau - \tilde{P}_\tau)|\mathcal{G}_t\}, \quad 0 \leq t \leq T, \qquad (7.2)$$

where $\mathcal{T}_{t,T}$ is the set of $\mathbb{G}$-stopping times taking values in $[t, T]$. Using repeated conditioning, we decompose (7.2) to

$$J_t = V_t - \tilde{P}_t,$$

where

$$V_t := \operatorname*{ess\,sup}_{\tau \in \mathcal{T}_{t,T}} \mathbb{E}^{\tilde{\mathbb{Q}}}\{e^{-\int_t^\tau r_v dv} P_\tau | \mathcal{G}_t\}. \qquad (7.3)$$

Hence, maximizing the price spread in (7.2) is equivalent to maximizing the expected discounted future market value P_τ under the investor's measure $\tilde{\mathbb{Q}}$ in (7.3).

The selection of the risk-neutral pricing measure $\tilde{\mathbb{Q}}$ can be based on the investor's hedging criterion or risk preferences. For instance, dynamic hedging under a quadratic criterion amounts to pricing under the well-known minimal martingale measure developed by Föllmer and Schweizer (1990). On the other hand, different risk-neutral pricing measures may also arise from marginal utility indifference pricing. In the cases of exponential and power utilities, this pricing mechanism will lead the investor to select the minimal entropy martingale measure (MEMM) (see Leung and Ludkovski (2012)) and the q-optimal martingale measure (see Henderson *et al.* (2005)).

Lemma 7.1. *For* $0 \leq t \leq T$, *we have* $V_t \geq P_t \vee \tilde{P}_t$. *Also,* $V_{\tau_d} = \tilde{P}_{\tau_d} = P_{\tau_d}$ *at default.*

Proof. Since $\tau = t$ and $\tau = T$ are candidate liquidation times, we conclude from (7.3) that $V_t \geq P_t \vee \tilde{P}_t$. Also, we observe from (7.1) that $P_t = \int_{(0,\tau_d]} e^{-\int_t^u r_v dv} dD_u = \tilde{P}_t$ for $t \geq \tau_d \wedge T$. This implies that

$$V_{\tau_d} = \operatorname*{ess\,sup}_{\tau \in \mathcal{T}_{\tau_d,T}} \mathbb{E}^{\tilde{\mathbb{Q}}}\{e^{-\int_{\tau_d}^{\tau} r_v dv} P_{\tau}|\mathcal{G}_{\tau_d}\}$$

$$= \operatorname*{ess\,sup}_{\tau \in \mathcal{T}_{\tau_d,T}} \mathbb{E}^{\tilde{\mathbb{Q}}}\{e^{-\int_{\tau_d}^{\tau} r_v dv} \tilde{P}_{\tau}|\mathcal{G}_{\tau_d}\} = \tilde{P}_{\tau_d} = P_{\tau_d}. \qquad (7.4)$$

$\square$

The last equation means that price discrepancy vanishes when the default event is observed or when the contract expires. This is also realistic since the market will no longer be liquid afterward.

If the defaultable claim is underpriced by the market at all times, that is, $P_t \leq \tilde{P}_t$, $\forall t \leq T$, then we infer from (7.2) that $J_t = 0$. This can be achieved at $\tau^* = T$ since price discrepancy must vanish at maturity, i.e. $P_T = \tilde{P}_T$. In turn, this implies that

$$V_t = \mathbb{E}^{\tilde{\mathbb{Q}}}\{e^{-\int_t^T r_v dv} P_T|\mathcal{G}_t\} = \mathbb{E}^{\tilde{\mathbb{Q}}}\{e^{-\int_t^T r_v dv} \tilde{P}_T|\mathcal{G}_t\} = \tilde{P}_t.$$

In this case, there is no benefit to liquidate before maturity T.

According to (7.3), the optimal liquidation timing directly depends on the investor's pricing measure $\tilde{\mathbb{Q}}$ as well as the market pricing measure $\mathbb{Q}$ (via the market price P). Specifically, we observe that the discounted market price $(e^{-\int_0^t r_v dv} P_t)_{0 \leq t \leq T}$ is a $(\mathbb{Q}, \mathbb{G})$-martingale, but generally not a $(\tilde{\mathbb{Q}}, \mathbb{G})$-martingale. If the discounted market price is a $(\tilde{\mathbb{Q}}, \mathbb{G})$-*supermartingale*, then it is optimal to sell the claim immediately. If the discounted market price turns out to be a $(\tilde{\mathbb{Q}}, \mathbb{G})$-*submartingale*, then it is optimal to delay the liquidation until maturity T. Besides these two scenarios, the optimal liquidation strategy may be non-trivial.

To quantify the value of optimally waiting to sell, we define the *delayed liquidation premium*:

$$L_t := V_t - P_t \geq 0. \qquad (7.5)$$

It is often more intuitive to study the optimal liquidation timing in terms of the premium L. Indeed, standard optimal stopping theory [Karatzas and

Shreve (1998), Appendix D] suggests that the optimal stopping time τ^* for (7.3) is the first time the process V reaches the reward P, namely,

$$\tau^* = \inf\{t \le u \le T : V_u = P_u\} = \inf\{t \le u \le T : L_u = 0\}. \quad (7.6)$$

The last equation, which follows directly from definition (7.5), implies that the investor will liquidate as soon as the delayed liquidation premium vanishes. Moreover, we observe from (7.4) and (7.6) that $\tau^* \le \tau_d$.

7.2 Optimal Liquidation under Markovian Credit Risk Models

We proceed to analyze the optimal liquidation problem under a general class of Markovian credit risk models. The description of various pricing measures will involve the mark-to-market risk premium and event risk premium, which are crucial in the characterization of the optimal liquidation strategy (see Theorem 7.5).

7.2.1 *Pricing Measures and Default Risk Premia*

We consider a n-dimensional Markovian state vector process $\mathbf{X}$ that drives the interest rate $r_t = r(t, \mathbf{X}_t)$ and default intensity $\hat{\lambda}_t = \hat{\lambda}(t, \mathbf{X}_t)$ for some positive measurable functions $r(\cdot, \cdot)$ and $\hat{\lambda}(\cdot, \cdot)$. Denote by $\mathbb{F}$ the filtration generated by $\mathbf{X}$. We also assume a Markovian payoff structure for the defaultable claim (Y, A, R, τ_d) with $Y = Y(\mathbf{X}_T)$, $A_t = \int_0^t q(u, \mathbf{X}_u)du$, and $R_t = R(t, \mathbf{X}_t)$ for some measurable functions $Y(\cdot)$, $q(\cdot, \cdot)$, and $R(\cdot, \cdot)$ satisfying integrability conditions (7.1.1).

Under the historical measure $\mathbb{P}$, the state vector process $\mathbf{X}$ satisfies the SDE

$$d\mathbf{X}_t = a(t, \mathbf{X}_t)dt + \Sigma(t, \mathbf{X}_t)d\mathbf{W}_t^{\mathbb{P}},$$

where $\mathbf{W}^{\mathbb{P}}$ is a m-dimensional $\mathbb{P}$-Brownian motion, a is the deterministic drift function, and Σ is the n by m deterministic volatility function.

Next, we consider the market pricing measure $\mathbb{Q} \sim \mathbb{P}$. To this end, we define the Radon-Nikodym density process $(Z_t^{\mathbb{Q},\mathbb{P}})_{0 \le t \le T}$ by

$$Z_t^{\mathbb{Q},\mathbb{P}} = \frac{d\mathbb{Q}}{d\mathbb{P}}\Big|\mathcal{G}_t = \mathcal{E}\big(-\phi^{\mathbb{Q},\mathbb{P}} \cdot \mathbf{W}^{\mathbb{P}}\big)_t \mathcal{E}\big((\mu - 1)M^{\mathbb{P}}\big)_t,$$

where the Doléans-Dade exponentials are defined by

$$\mathcal{E}\big(-\boldsymbol{\phi}^{\mathbb{Q},\mathbb{P}}\cdot\mathbf{W}^{\mathbb{P}}\big)_t := \exp\bigg(-\frac{1}{2}\int_0^t \|\boldsymbol{\phi}_u^{\mathbb{Q},\mathbb{P}}\|^2 du - \int_0^t \boldsymbol{\phi}_u^{\mathbb{Q},\mathbb{P}}\cdot d\mathbf{W}_u^{\mathbb{P}}\bigg), \qquad (7.7)$$

$$\mathcal{E}\big((\mu-1)M^{\mathbb{P}}\big)_t := \exp\bigg(\int_0^t \log(\mu_{u-})dN_u - \int_0^t (1-N_u)(\mu_u-1)\hat{\lambda}_u du\bigg),$$
$$(7.8)$$

and $M_t^{\mathbb{P}} := N_t - \int_0^t (1-N_u)\hat{\lambda}_u du$ is the compensated $(\mathbb{P},\mathbb{G})$-martingale associated with N. Here, $(\boldsymbol{\phi}_t^{\mathbb{Q},\mathbb{P}})_{0\le t\le T}$ and $(\mu_t)_{0\le t\le T}$ are adapted processes satisfying $\int_0^T \|\boldsymbol{\phi}_u^{\mathbb{Q},\mathbb{P}}\|^2 du < \infty$, $\mu \ge 0$, and $\int_0^T \mu_u\hat{\lambda}_u du < \infty$ (see Theorem 4.8 of Schönbucher (2003)).

The process $\boldsymbol{\phi}^{\mathbb{Q},\mathbb{P}}$ is commonly referred to as the *mark-to-market risk premium*, which is assumed herein to be Markovian of the form $\boldsymbol{\phi}^{\mathbb{Q},\mathbb{P}}(t,\mathbf{X}_t)$. The process μ is referred to as *event risk premium*, which captures the compensation from the uncertain timing of default. The $\mathbb{Q}$-default intensity, denoted by λ, is related to $\mathbb{P}$-intensity via $\lambda_t = \mu_t\hat{\lambda}_t$. Here, we also assume μ to be Markovian of the form $\mu(t,\mathbf{X}_t) = \lambda(t,\mathbf{X}_t)/\hat{\lambda}(t,\mathbf{X}_t)$.

By multi-dimensional Girsanov Theorem, it follows that $\mathbf{W}_t^{\mathbb{Q}} := \mathbf{W}_t^{\mathbb{P}} + \int_0^t \boldsymbol{\phi}_u^{\mathbb{Q},\mathbb{P}}du$ is a m-dimensional $\mathbb{Q}$-Brownian motion, and $M_t^{\mathbb{Q}} := N_t - \int_0^t (1-N_u)\mu_u\hat{\lambda}_u du$ is a $(\mathbb{Q},\mathbb{G})$-martingale. Consequently, the $\mathbb{Q}$-dynamics of $\mathbf{X}$ are given by

$$d\mathbf{X}_t = b(t,\mathbf{X}_t)dt + \Sigma(t,\mathbf{X}_t)d\mathbf{W}_t^{\mathbb{Q}},$$

where $b(t,\mathbf{X}_t) := a(t,\mathbf{X}_t) - \Sigma(t,\mathbf{X}_t)\boldsymbol{\phi}^{\mathbb{Q},\mathbb{P}}(t,\mathbf{X}_t)$.

Similarly, the investor's pricing measure $\tilde{\mathbb{Q}}$ is related to the historical measure $\mathbb{P}$ through the investor's Markovian risk premium functions $\boldsymbol{\phi}^{\tilde{\mathbb{Q}},\mathbb{P}}(t,\mathbf{x})$ and $\tilde{\mu}(t,\mathbf{x})$. Precisely, the measure $\tilde{\mathbb{Q}}$ is defined by the density process $Z_t^{\tilde{\mathbb{Q}},\mathbb{P}} = \mathcal{E}\big(-\boldsymbol{\phi}^{\tilde{\mathbb{Q}},\mathbb{P}}\cdot\mathbf{W}^{\mathbb{P}}\big)_t \mathcal{E}\big((\tilde{\mu}-1)M^{\mathbb{P}}\big)_t$. By a change of measure, the drift of $\mathbf{X}$ under $\tilde{\mathbb{Q}}$ is modified to $\tilde{b}(t,\mathbf{X}_t) := a(t,\mathbf{X}_t) - \Sigma(t,\mathbf{X}_t)\boldsymbol{\phi}^{\tilde{\mathbb{Q}},\mathbb{P}}(t,\mathbf{X}_t)$.

Then, the EMMs $\mathbb{Q}$ and $\tilde{\mathbb{Q}}$ are related by the Radon-Nikodym derivative:

$$Z_t^{\tilde{\mathbb{Q}},\mathbb{Q}} = \frac{d\tilde{\mathbb{Q}}}{d\mathbb{Q}}\Big|\mathcal{G}_t = \mathcal{E}\big(-\boldsymbol{\phi}^{\tilde{\mathbb{Q}},\mathbb{Q}}\cdot\mathbf{W}^{\mathbb{Q}}\big)_t \mathcal{E}\big((\frac{\tilde{\mu}}{\mu}-1)M^{\mathbb{Q}}\big)_t,$$

where the Doléans-Dade exponentials are defined by

$$\mathcal{E}\big(-\boldsymbol{\phi}^{\tilde{\mathbb{Q}},\mathbb{Q}}\cdot\mathbf{W}^{\mathbb{Q}}\big)_t := \exp\bigg(-\frac{1}{2}\int_0^t \|\boldsymbol{\phi}_u^{\tilde{\mathbb{Q}},\mathbb{Q}}\|^2 du - \int_0^t \boldsymbol{\phi}_u^{\tilde{\mathbb{Q}},\mathbb{Q}}\cdot d\mathbf{W}_u^{\mathbb{Q}}\bigg), \qquad (7.9)$$

$$\mathcal{E}\big((\frac{\tilde{\mu}}{\mu}-1)M^{\mathbb{Q}}\big)_t := \exp\bigg(\int_0^t \log(\frac{\tilde{\mu}_{u-}}{\mu_{u-}})dN_u - \int_0^t (1-N_u)(\frac{\tilde{\mu}_u}{\mu_u}-1)\lambda_u du\bigg).$$
$$(7.10)$$

We observe that $\phi_t^{\tilde{\mathbb{Q}},\mathbb{Q}} = \phi_t^{\tilde{\mathbb{Q}},\mathbb{P}} - \phi_t^{\mathbb{Q},\mathbb{P}}$ from the decomposition:

$$\phi_t^{\tilde{\mathbb{Q}},\mathbb{Q}} dt = d\mathbf{W}_t^{\tilde{\mathbb{Q}}} - d\mathbf{W}_t^{\mathbb{Q}} = (d\mathbf{W}_t^{\tilde{\mathbb{Q}}} - d\mathbf{W}_t^{\mathbb{P}}) - (d\mathbf{W}_t^{\mathbb{Q}} - d\mathbf{W}_t^{\mathbb{P}})$$

$$= (\phi_t^{\tilde{\mathbb{Q}},\mathbb{P}} - \phi_t^{\mathbb{Q},\mathbb{P}}) dt. \tag{7.11}$$

Therefore, we can interpret $\phi_t^{\tilde{\mathbb{Q}},\mathbb{Q}}$ as the incremental mark-to-market risk premium assigned by the investor relative to the market. On the other hand, the discrepancy in event risk premia is accounted for in the second Doléans-Dade exponential (7.10).

Example 7.2. *The OU Model.* Suppose $(r, \hat{\lambda}) = \mathbf{X}$, following the OU dynamics:

$$\begin{pmatrix} dr_t \\ d\hat{\lambda}_t \end{pmatrix} = \begin{pmatrix} \hat{\kappa}_r(\hat{\theta}_r - r_t) \\ \hat{\kappa}_\lambda(\hat{\theta}_\lambda - \hat{\lambda}_t) \end{pmatrix} dt + \begin{pmatrix} \sigma_r & 0 \\ \sigma_\lambda \rho & \sigma_\lambda\sqrt{1-\rho^2} \end{pmatrix} \begin{pmatrix} dW_t^{1,\mathbb{P}} \\ dW_t^{2,\mathbb{P}} \end{pmatrix},$$

with constant parameters $\hat{\kappa}_r, \hat{\theta}_r, \hat{\kappa}_\lambda, \hat{\theta}_\lambda \geq 0$. Here, $\hat{\kappa}_r$, $\hat{\kappa}_\lambda$ parametrize the speed of mean reversion, and $\hat{\theta}_r$, $\hat{\theta}_\lambda$ represent the long-term means (see Section 7.1 of Schönbucher (2003)). Assuming a constant event risk premium μ by the market, the $\mathbb{Q}$-intensity is specified by $\lambda_t = \mu\hat{\lambda}_t$ and the pair (r, λ) satisfies SDEs:

$$\begin{pmatrix} dr_t \\ d\lambda_t \end{pmatrix} = \begin{pmatrix} \kappa_r(\theta_r - r_t) \\ \kappa_\lambda(\mu\theta_\lambda - \lambda_t) \end{pmatrix} dt + \begin{pmatrix} \sigma_r & 0 \\ \mu\sigma_\lambda\rho & \mu\sigma_\lambda\sqrt{1-\rho^2} \end{pmatrix} \begin{pmatrix} dW_t^{1,\mathbb{Q}} \\ dW_t^{2,\mathbb{Q}} \end{pmatrix},$$

with constants $\kappa_r, \theta_r, \kappa_\lambda, \theta_\lambda \geq 0$. Under the investor's measure $\tilde{\mathbb{Q}}$, the SDEs for r_t and $\tilde{\lambda}_t = \tilde{\mu}\hat{\lambda}_t$ are of the same form with parameters $\tilde{\kappa}_r, \tilde{\theta}_r, \tilde{\kappa}_\lambda, \tilde{\theta}_\lambda$ and $\tilde{\mu}$, and $\mathbf{W}^{\mathbb{Q}}$ is replaced by $\mathbf{W}^{\tilde{\mathbb{Q}}}$.

Direct computation yields the relative mark-to-market risk premium:

$$\phi_t^{\tilde{\mathbb{Q}},\mathbb{Q}} = \begin{pmatrix} \frac{\kappa_r(\theta_r - r_t) - \tilde{\kappa}_r(\tilde{\theta}_r - r_t)}{\sigma_r} \\ \frac{1}{\sqrt{1-\rho^2}} \frac{\kappa_\lambda(\theta_\lambda - \hat{\lambda}_t) - \tilde{\kappa}_\lambda(\tilde{\theta}_\lambda - \hat{\lambda}_t)}{\sigma_\lambda} - \frac{\rho}{\sqrt{1-\rho^2}} \frac{\kappa_r(\theta_r - r_t) - \tilde{\kappa}_r(\tilde{\theta}_r - r_t)}{\sigma_r} \end{pmatrix}.$$

The upper term is the incremental risk premium for the interest rate while the bottom term reflects the discrepancy in the default risk premia (see (7.11)).

Example 7.3. *The CIR Model.* Let $\mathbf{X} = (X^1, \ldots, X^n)^T$ follow the multi-factor CIR model:[2]

$$dX_t^i = \hat{\kappa}_i(\hat{\theta}_i - X_t^i)dt + \sigma_i\sqrt{X_t^i}\,dW_t^{i,\mathbb{P}},$$

---

[2] See Section 7.2 of Schönbucher (2003) for details.

where $W^{i,\mathbb{P}}$ are mutually independent $\mathbb{P}$-Brownian motions and $\hat{\kappa}_i$, $\hat{\theta}_i$, $\sigma_i \geq 0$, $i = 1, \ldots, n$ satisfy Feller condition $2\hat{\kappa}_i\hat{\theta}_i > \sigma_i^2$. The interest rate r and historical default intensity $\hat{\lambda}$ are non-negative linear combinations of X^i with constant weights $w_i^r, w_i^\lambda \geq 0$, namely, $r_t = \sum_{i=1}^{n} w_i^r X_t^i$ and $\hat{\lambda}_t = \sum_{i=1}^{n} w_i^\lambda X_t^i$. Under measure $\mathbb{Q}$, X^i satisfies the SDE:

$$dX_t^i = \kappa_i(\theta_i - X_t^i)dt + \sigma_i\sqrt{X_t^i}\,dW_t^{i,\mathbb{Q}},$$

with new mean reversion speed κ_i and long-run mean θ_i.

Under the investor's measure $\tilde{\mathbb{Q}}$, the SDE for the state vector is of the same form with new parameters $\tilde{\kappa}_i$, $\tilde{\theta}_i$. The associated relative mark-to-market risk premium has following structure:

$$\phi_{i,t}^{\tilde{\mathbb{Q}},\mathbb{Q}} = \frac{\kappa_i(\theta_i - X_t^i) - \tilde{\kappa}_i(\tilde{\theta}_i - X_t^i)}{\sigma_i\sqrt{X_t^i}}.$$

The event risk premia $(\mu, \tilde{\mu})$ are assigned via $\lambda_t = \mu\hat{\lambda}_t$ under $\mathbb{Q}$ and $\tilde{\lambda}_t = \tilde{\mu}\hat{\lambda}_t$ under $\tilde{\mathbb{Q}}$ respectively.

Remark 7.4. The current framework can be readily generalized to the situation where the investor needs to assume an alternative historical measure $\tilde{\mathbb{P}}$. The resulting risk premium $\phi^{\tilde{\mathbb{Q}},\mathbb{Q}}$ will have a third decomposition component $\phi^{\tilde{\mathbb{P}},\mathbb{P}}$, reflecting the difference in historical dynamics.

For any defaultable claim (Y, A, R, τ_d), the ex-dividend pre-default market price is given by

$$\begin{aligned}
C(t, \mathbf{X}_t) = \mathbb{E}^{\mathbb{Q}}\Big\{ & e^{-\int_t^T (r_v + \lambda_v)dv} Y(\mathbf{X}_T) \\
& + \int_t^T e^{-\int_t^u (r_v + \lambda_v)dv} \big(\lambda_u R(u, \mathbf{X}_u) + q(u, \mathbf{X}_u)\big) du \Big| \mathcal{F}_t \Big\}
\end{aligned} \quad (7.12)$$

The associated cumulative price is related to the pre-default price via

$$\begin{aligned}
P_t = &(1 - N_t)C(t, \mathbf{X}_t) + \int_0^t (1 - N_u)q(u, \mathbf{X}_u)e^{\int_u^t r_v dv} du \\
& + \int_{(0,t]} R(u, \mathbf{X}_u)e^{\int_u^t r_v dv} dN_u.
\end{aligned}$$

The price function $C(t, \mathbf{x})$ can be determined by solving the PDE:

$$\begin{cases}
\dfrac{\partial C}{\partial t}(t, \mathbf{x}) + \mathcal{L}_{b,\lambda}C(t, \mathbf{x}) + \lambda(t, \mathbf{x})R(t, \mathbf{x}) + q(t, \mathbf{x}) = 0, & (t, \mathbf{x}) \in [0, T) \times \mathbb{R}^n, \\
C(T, \mathbf{x}) = Y(\mathbf{x}), & \mathbf{x} \in \mathbb{R}^n,
\end{cases}$$

$$(7.13)$$

where $\mathcal{L}_x$ is the operator defined by

$$\mathcal{L}_{b,\lambda}f = \sum_{i=1}^{n} b_i(t,\mathbf{x})\frac{\partial f}{\partial x_i} + \frac{1}{2}\sum_{i,j=1}^{n}(\Sigma(t,\mathbf{x})\Sigma(t,\mathbf{x})^T)_{ij}\frac{\partial^2 f}{\partial x_i \partial x_j}$$
$$- \big(r(t,\mathbf{x}) + \lambda(t,\mathbf{x})\big)f. \tag{7.14}$$

The computation is similar for the investor's price under $\tilde{\mathbb{Q}}$.

7.2.2 Delayed Liquidation Premium and Optimal Timing

Next, we analyze the optimal liquidation problem V defined in (7.3) for the general defaultable claim under the current Markovian setting.

Theorem 7.5. *For a general defaultable claim* (Y, A, R, τ_d) *under the Markovian credit risk model, the delayed liquidation premium admits the probabilistic representation:*

$$L_t = \mathbf{1}_{\{t < \tau_d\}} \operatorname*{ess\,sup}_{\tau \in \mathcal{T}_{t,T}} \mathbb{E}^{\tilde{\mathbb{Q}}}\Big\{ \int_t^\tau e^{-\int_t^u (r_v + \tilde{\lambda}_v)dv} G(u, \mathbf{X}_u)du \Big| \mathcal{F}_t \Big\}, \tag{7.15}$$

where $G : [0, T] \times \mathbb{R}^n \mapsto \mathbb{R}$ *is defined by*

$$G(t,\mathbf{x}) = -\big(\nabla_x C(t,\mathbf{x})\big)^T \Sigma(t,\mathbf{x})\phi^{\tilde{\mathbb{Q}},\mathbb{Q}}(t,\mathbf{x})$$
$$+ \big(R(t,\mathbf{x}) - C(t,\mathbf{x})\big)\big(\tilde{\mu}(t,\mathbf{x}) - \mu(t,\mathbf{x})\big)\hat{\lambda}(t,\mathbf{x}). \tag{7.16}$$

If $G(t,\mathbf{x}) \geq 0 \ \forall(t,\mathbf{x})$, *then it is optimal to delay the liquidation till maturity* T.
If $G(t,\mathbf{x}) \leq 0 \ \forall(t,\mathbf{x})$, *then it is optimal to sell immediately.*

Proof. First, we look at the $\tilde{\mathbb{Q}}$-dynamics of discounted market price $(e^{-\int_t^u r_v dv}P_u)_{t \leq u \leq T}$. Applying Corollary 2.2 of Bielecki *et al.* (2008), for $t \leq u \leq T$,

$$d(e^{-\int_t^u r_v dv}P_u) = e^{-\int_t^u r_v dv}[(R_u - C_u)dM_u^{\mathbb{Q}} + (1 - N_u)(\nabla_x C_u)^T \Sigma_u d\mathbf{W}_u^{\mathbb{Q}}] \tag{7.17}$$

$$= e^{-\int_t^u r_v dv}\big((1 - N_u)G(u, \mathbf{X}_u)du$$
$$+ (1 - N_u)(\nabla_x C_u)^T \Sigma_u d\mathbf{W}_u^{\tilde{\mathbb{Q}}} + (R_u - C_u)dM_u^{\tilde{\mathbb{Q}}}\big),$$

where G is defined in (7.16), and $M^{\tilde{\mathbb{Q}}}$ is the compensated $(\tilde{\mathbb{Q}}, \mathbb{G})$-martingale for N. Consequently,

$$L_t = \operatorname*{ess\,sup}_{\tau \in \mathcal{T}_{t,T}} \mathbb{E}^{\tilde{\mathbb{Q}}}\Big\{ \int_t^\tau (1 - N_u)e^{-\int_t^u r_v dv} G(u, \mathbf{X}_u)du \Big| \mathcal{G}_t \Big\},$$

where (7.15) follows from the change of filtration technique [Bielecki and Rutkowski (2002), §5.1.1]. If $G \geq 0$, then the integrand in (7.15) is positive a.s. and therefore the largest possible stopping time T is optimal. If $G \leq 0$, then $\tau^* = t$ is optimal and $L_t = 0$ a.s.   □

The drift function G has two components explicitly depending on $\phi^{\tilde{Q},Q}$ and $\tilde{\mu} - \mu$. If $\phi^{\tilde{Q},Q}(t, \mathbf{x}) = \mathbf{0} \; \forall(t, \mathbf{x})$, that is, the investor and market agree on the mark-to-market risk premium, then the sign of G is solely determined by the difference $\tilde{\mu} - \mu$, since recovery R in general is less than the pre-default price C. On the other hand, if $\mu(t, \mathbf{x}) = \tilde{\mu}(t, \mathbf{x}) \; \forall(t, \mathbf{x})$, then the second term of G vanishes but G still depends on μ through $\nabla_x C$ in the first term.

Theorem 7.5 allows us to conclude the optimal liquidation timing when the drift function is of constant sign. In other cases, the optimal liquidation policy may be non-trivial and needs to be numerically determined. For this purpose, we write $L_t = \mathbf{1}_{\{t < \tau_d\}} \hat{L}(t, \mathbf{X}_t)$, where $\hat{L}$ is the (Markovian) pre-default delayed liquidation premium defined by

$$\hat{L}(t, \mathbf{X}_t) = \operatorname*{ess\,sup}_{\tau \in \mathcal{T}_{t,T}} \mathbb{E}^{\tilde{Q}}\Big\{ \int_t^\tau e^{-\int_t^u (r_v + \tilde{\lambda}_v)\,dv} G(u, \mathbf{X}_u)\,du \,\big|\, \mathcal{F}_t \Big\}. \quad (7.18)$$

We determine $\hat{L}$ from the variational inequality :

$$\min\Big(-\frac{\partial \hat{L}}{\partial t}(t, \mathbf{x}) - \mathcal{L}_{\tilde{b},\tilde{\lambda}} \hat{L}(t, \mathbf{x}) - G(t, \mathbf{x}), \; \hat{L}(t, \mathbf{x}) \Big) = 0, \quad (7.19)$$

for $(t, \mathbf{x}) \in [0, T) \times \mathbb{R}^n$, where $\mathcal{L}_{\tilde{b},\tilde{\lambda}}$ is defined in (7.14), and the terminal condition is $\hat{L}(T, \mathbf{x}) = 0$, for $\mathbf{x} \in \mathbb{R}^n$.

The investor's optimal timing is characterized by the sell region $\mathcal{S}$ and delay region $\mathcal{D}$, namely,

$$\mathcal{S} = \{(t, \mathbf{x}) \in [0, T] \times \mathbb{R}^n : \hat{L}(t, \mathbf{x}) = 0\}, \quad (7.20)$$

$$\mathcal{D} = \{(t, \mathbf{x}) \in [0, T] \times \mathbb{R}^n : \hat{L}(t, \mathbf{x}) > 0\}. \quad (7.21)$$

Also, define $\hat{\tau}^* = \inf\{t \leq u \leq T : \hat{L}_u = 0\}$. On $\{\hat{\tau}^* \geq \tau_d\}$, liquidation occurs at τ_d since $L_{\tau_d} = 0$. On $\{\hat{\tau}^* < \tau_d\}$, $\hat{\tau}^*$ is optimal since when $u < \hat{\tau}^*$, $L_u = \mathbf{1}_{\{u < \tau_d\}} \hat{L}_u > 0$ and $L_{\hat{\tau}^*} = 0$. Incorporating the observation of τ_d, the optimal stopping time is $\tau^* = \hat{\tau}^* \wedge \tau_d$.

Hence, given no default by time t and $\mathbf{X}_t = \mathbf{x}$, it is optimal to wait at the current time t if $\hat{L}(t, \mathbf{x}) > 0$ in view of the delay region $\mathcal{D}$ in (7.21). This is also intuitive as there is a strictly positive premium for delaying liquidation. On the other hand, the sell region $\mathcal{S}$ must lie within the set

$G_- := \{(t, \mathbf{x}) : G(t, \mathbf{x}) \le 0\}$. To see this, we infer from (7.18) that, for any given point $(t, \mathbf{x})$ such that $\hat{L}(t, \mathbf{x}) = 0$, we must have $G(t, \mathbf{x}) \le 0$. In turn, the delay region $\mathcal{D}$ must contain the set $G_+ := \{(t, \mathbf{x}) : G(t, \mathbf{x}) > 0\}$. From these observations, one can obtain some insights about the sell and delay regions by inspecting $G(t, \mathbf{x})$, which is much easier to compute than $\hat{L}(t, \mathbf{x})$. We shall illustrate this numerically below.

Lastly, let us consider a special example where the stochastic factor $\mathbf{X}$ is absent from the model. With reference to (7.12), we set a constant terminal payoff Y, and deterministic recovery $R(t)$ and coupon rate $q(t)$. Suppose the investor and market perceive the same deterministic interest rate $r(t)$, but possibly different deterministic default intensities, respectively, $\tilde{\lambda}(t) = \tilde{\mu}(t)\hat{\lambda}(t)$ and $\lambda(t) = \mu(t)\hat{\lambda}(t)$. In this case, the price function C in (7.16) will depend only on t but not on $\mathbf{x}$, and there will be no mark-to-market risk premium. Therefore, the first term of drift function in (7.16) will vanish. However, the second term remains due to potential discrepancy in event risk premium, i.e. $\tilde{\mu}(t) \ne \mu(t)$. As a result, the drift function reduces to

$$G(t) = (R(t) - C(t))(\tilde{\mu}(t) - \mu(t))\hat{\lambda}(t).$$

Furthermore, the absence of the stochastic factor $\mathbf{X}$ also trivializes the filtration $\mathbb{F}$, and leads the investor to optimize over only constant times. The delayed liquidation premium admits the form: $L_t = \mathbf{1}_{\{t < \tau_d\}}\hat{L}(t)$, where $\hat{L}(t)$ is a deterministic function given by

$$\hat{L}(t) = \sup_{t \le \hat{t} \le T} \int_t^{\hat{t}} e^{-\int_t^u (r(v) + \tilde{\lambda}(v))dv} G(u)du. \tag{7.22}$$

As in Theorem 7.5, if G is always positive (resp. negative) over $[t, T]$, then the optimal time $\hat{t}^* = T$ (resp. $\hat{t}^* = t$). Otherwise, differentiating the integral in (7.22) implies that the deterministic candidate times also include the roots of $G(\hat{t}) = 0$. Therefore, we select among the candidate times t, T and the roots of G to see which would yield the largest integral value in (7.22).

7.3   Application to Single-Name Credit Derivatives

We proceed to illustrate our analysis for a number of credit derivatives, with an emphasis on how risk premia discrepancy affects the optimal liquidation strategies.

7.3.1 Defaultable Bonds with Zero Recovery

Consider a defaultable zero-coupon zero-recovery bond with face value 1 and maturity T. By a change of filtration, the market price of the zero-coupon zero-recovery bond is given by

$$P_t^0 := \mathbb{E}^{\mathbb{Q}}\{e^{-\int_t^T r_v dv} \mathbf{1}_{\{\tau_d > T\}} | \mathcal{G}_t\}$$

$$= \mathbf{1}_{\{t < \tau_d\}} \mathbb{E}^{\mathbb{Q}}\{e^{-\int_t^T (r_v + \lambda_v) dv} | \mathcal{F}_t\} = \mathbf{1}_{\{t < \tau_d\}} C^0(t, \mathbf{X}_t),$$

where C^0 denotes the market pre-default price that solves (7.13). Under the general Markovian credit risk model in Section 7.2.1, we can apply Theorem 7.5 with the quadruple $(1, 0, 0, \tau_d)$ to obtain the corresponding drift function.

Under the OU dynamics in Section 7.2, the pre-default price function $C^0(t, r, \lambda)$ is given explicitly by

$$C^0(t, r, \lambda) = e^{A(T-t) - B(T-t)r - D(T-t)\lambda},$$

where

$$B(s) = \frac{1 - e^{-\kappa_r s}}{\kappa_r}, \quad D(s) = \frac{1 - e^{-\kappa_\lambda s}}{\kappa_\lambda},$$

$$A(s) = \int_0^s [\frac{1}{2}\sigma_r^2 B^2(z) + \rho\mu\sigma_r\sigma_\lambda B(z)D(z) + \frac{1}{2}\mu^2\sigma_\lambda^2 D^2(z)$$

$$- \kappa_r\theta_r B(z) - \mu\kappa_\lambda\theta_\lambda D(z)]dz.$$

As a result, the drift function $G^0(t, r, \lambda)$ admits a separable form:

$$G^0(t, r, \lambda) = C^0(t, r, \lambda)\Big(B(T-t)(\tilde{\kappa}_r - \kappa_r)r + B(T-t)(\kappa_r\theta_r - \tilde{\kappa}_r\tilde{\theta}_r)$$

$$+ [D(T-t)(\tilde{\kappa}_\lambda - \kappa_\lambda) - (\frac{\tilde{\mu}}{\mu} - 1)]\lambda + \mu D(T-t)(\kappa_\lambda\theta_\lambda - \tilde{\kappa}_\lambda\tilde{\theta}_\lambda)\Big).$$

We can draw several insights on the liquidation timing from this drift function. If the market and the investor agree on the speed of mean reversion for interest rate, i.e. $\kappa_r = \tilde{\kappa}_r$, then $G^0(t, r, \lambda)/C^0(t, r, \lambda)$ is linear in λ. Furthermore, if the slope $D(T-t)(\tilde{\kappa}_\lambda - \kappa_\lambda) - (\frac{\tilde{\mu}}{\mu} - 1)$ and intercept $B(T-t)(\kappa_r\theta_r - \tilde{\kappa}_r\tilde{\theta}_r) + \mu D(T-t)(\kappa_\lambda\theta_\lambda - \tilde{\kappa}_\lambda\tilde{\theta}_\lambda)$ are of the same sign, then the optimal liquidation strategy must be trivial in view of Theorem 7.5. In contrast, if the slope and intercept differ in signs, the optimal stopping problem may be nontrivial and the sign of the slope determines qualitative properties of optimal stopping rules. For instance, suppose the slope is positive. We infer that it is optimal for the holder to wait at high default intensity where the corresponding G^0 and thus delayed liquidation premium are positive. The converse holds if the slope is negative.

If the investor disagrees with market only on event risk premium, i.e. $\mu \neq \tilde{\mu}$, then the drift function is reduced to $G^0(t, r, \lambda) = -C^0(t, r, \lambda)(\frac{\tilde{\mu}}{\mu} - 1)\lambda$, which is of constant sign. This implies trivial strategies. If $\mu > \tilde{\mu}$, then $G^0 > 0$ and it is optimal to delay the liquidation until maturity. On the other hand, if $\mu < \tilde{\mu}$, then it is optimal to sell immediately. More general specifications of the event risk premium could depend on the state vector and may lead to nontrivial optimal stopping rules. Disagreement on mean level θ_λ has a similar effect to that of μ.

If the investor disagrees with market only on speed of mean reversion, i.e. $\kappa_\lambda \neq \tilde{\kappa}_\lambda$, then $G^0(t, r, \lambda) = C^0(t, r, \lambda)D(T - t)\big[(\tilde{\kappa}_\lambda - \kappa_\lambda)\lambda + \mu\theta_\lambda(\kappa_\lambda - \tilde{\kappa}_\lambda)\big]$ with $D(T - t) > 0$ before T, where the slope and intercept differ in signs. If $\kappa_\lambda < \tilde{\kappa}_\lambda$, the slope $\tilde{\kappa}_\lambda - \kappa_\lambda$ is positive and it is optimal to sell immediately at a low intensity, and thus, a high bond price. The converse holds for $\kappa_\lambda > \tilde{\kappa}_\lambda$.

We consider a numerical example where the interest rate is constant and the market default intensity λ is chosen as the state vector $\mathbf{X}$ with OU dynamics. We employ the standard implicit PSOR algorithm to solve $\hat{L}(t, \lambda)$ through its variational inequality (7.19) over a uniform finite grid with Neumann condition applied on the intensity boundary. The market parameters are $T = 1$, $\mu = 2$, $\kappa_\lambda = 0.2$, $\theta_\lambda = 0.015$, $r = 0.03$, and $\sigma = 0.02$.[3]

From formula (7.3.1), we observe a one-to-one correspondence between the market pre-default bond price C^0 and its default intensity λ for any fixed (t, r), namely,

$$\lambda = \frac{-\log(C^0) + A(T - t) - B(T - t)r}{D(T - t)}. \tag{7.23}$$

Substituting (7.23) into (7.20) and (7.21), we can characterize the sell region and delay region in terms of the observable pre-default market price C^0.

---

[3]These values are based on the estimates in Driessen (2005); Duffee (1999).

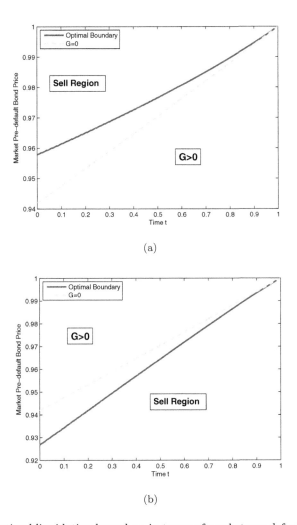

Fig. 7.1 Optimal liquidation boundary in terms of market pre-default bond price under OU dynamics. We take $T = 1$, $r = 0.03$, $\sigma = 0.02$, $\mu = \tilde{\mu} = 2$, and $\theta_\lambda = \tilde{\theta}_\lambda = 0.015$. Panel (a): When $\kappa_\lambda = 0.2 < 0.3 = \tilde{\kappa}_\lambda$, the optimal boundary increases from 0.958 to 1 over time. Panel (b): When $\kappa_\lambda = 0.3 > 0.2 = \tilde{\kappa}_\lambda$, the optimal boundary increases from 0.927 to 1 over time. The dashed straight line is defined by $G = 0$, and we have $G \leq 0$ in both sell regions.

In Figure 7.1(a), we assume that the investor agrees with the market on all parameters, but has a higher speed of mean reversion $\tilde{\kappa}_\lambda > \kappa_\lambda$. In this case, the investor tends to sell the bond at a high market price, which is consistent with our previous analysis in terms of drift function. If the bond price starts below 0.958 at time 0, the optimal liquidation strategy for the investor is to hold and sell the bond as soon as the price hits the optimal boundary. If the bond price starts above 0.958 at time 0, the optimal liquidation strategy is to sell immediately. In the opposite case where $\tilde{\kappa}_\lambda < \kappa_\lambda$ (see Figure 7.1(b)), the optimal liquidation strategy is reversed – it is optimal to sell at a lower boundary. In each cases, the sell region must lie within where G is non-positive, and the straight line defined by $G = 0$ can be viewed as a linear approximation of the optimal liquidation boundary.

Under the CIR dynamics in Section 7.2, C^0 admits a closed-form formula:

$$C^0(t, \mathbf{x}) = \prod_{i=1}^{n} \mathbb{E}^{\mathbb{Q}}\{e^{-\int_t^T (w_i^r + \mu w_i^\lambda) X_v^i dv} X_v^i dv | \mathbf{X}_t = \mathbf{x}\} = \prod_{i=1}^{n} A_i(T - t) e^{-B_i(T-t)x_i},$$

where

$$A_i(s) = [\frac{2\Xi_i e^{(\Xi_i + \kappa_i)s/2}}{(\Xi_i + \kappa_i)(e^{\Xi_i s} - 1) + 2\Xi_i}]^{2\kappa_i\theta_i/\sigma_i^2},$$

$$B_i(s) = \frac{2(e^{\Xi_i s} - 1)(w_i^r + \mu w_i^\lambda)}{(\Xi_i + \kappa_i)(e^{\Xi_i s} - 1) + 2\Xi_i}, \quad \text{and} \quad \Xi_i = \sqrt{\kappa_i^2 + 2\sigma_i^2(w_i^r + \mu w_i^\lambda)}.$$

As a result, the drift function is given by

$$G^0(t, \mathbf{x}) = [\sum_{i=1}^{n} ([B_i(T - t)(\tilde{\kappa}_i - \kappa_i) - (\tilde{\mu} - \mu)w_i^\lambda]x_i$$
$$+ B_i(T - t)(\kappa_i\theta_i - \tilde{\kappa}_i\tilde{\theta}_i))]C^0(t, \mathbf{x}),$$

which is again linear in terms of $C^0(t, \mathbf{x})$.

To illustrate the optimal liquidation strategy, we consider a numerical example where interest rate is constant, $\mathbf{X} = \lambda$, and $\mathbf{w}^\lambda = \frac{1}{\mu}$. The benchmark specifications for the market default intensity λ in the CIR dynamics are $T = 1$, $\mu = 2$, $\kappa_\lambda = 0.2$, $\theta_\lambda = 0.015$, $r = 0.03$, and $\sigma = 0.07$. Like in the OU model, we can again express the sell region and delay region in terms of the pre-default market price C^0; see Figure 7.2.

7.3.2 *Recovery of Treasury and Market Value*

Extending the preceding analysis on defaultable bonds, we incorporate two principle ways of modeling recovery: the recovery of treasury or market value.

By the recovery of treasury, we assume that a recovery of c times the value of the equivalent default-free bond is paid upon default. Therefore, the market pre-default bond price function is

$$C^{RT}(t, \mathbf{x}) = (1 - c)C^0(t, \mathbf{x}) + c\beta(t, \mathbf{x}),$$

where $\beta(t, \mathbf{x}) := \mathbb{E}^{\mathbb{Q}}\{e^{-\int_t^T r_v dv} | \mathbf{X}_t = \mathbf{x}\}$ is the equivalent default-free bond price. Then, applying Theorem 7.5 with the quadruple $(1, 0, c\beta, \tau_d)$, we obtain the corresponding drift function:

$$G^{RT}(t, \mathbf{x}) = -\left(\nabla_x C^{RT}(t, \mathbf{x})\right)^T \Sigma(t, \mathbf{x}) \phi^{\tilde{Q}, Q}(t, \mathbf{x})$$
$$+ (c - 1)\left(\tilde{\mu}(t, \mathbf{x}) - \mu(t, \mathbf{x})\right)\hat{\lambda}(t, \mathbf{x})C^0(t, \mathbf{x}). \quad (7.24)$$

If $c = 0$, then $C^{RT}(t, \mathbf{x}) = C^0(t, \mathbf{x})$ and G^{RT} in (7.24) reduces to the drift function of the zero-recovery bond. If $c = 1$, then $C^{RT}(t, \mathbf{x}) = \beta(t, \mathbf{x})$ is the market price of a default-free bond, and risk premium discrepancy may arise only from the interest rate dynamics.

Here are two examples where the drift function G^{RT} in (7.24) can be computed explicitly.

Example 7.6. Under OU model, $C^{RT}(t, r, \lambda)$ is computed according to (7.3.2) with $C^0(t, r, \lambda)$ in (7.3.1) and $\beta(t, r, \lambda) = e^{\bar{A}(T-t)-B(T-t)r}$, where $\bar{A}(s) = \int_0^s \left[\frac{1}{2}\sigma_r^2 B^2(z) - \kappa_r \theta_r B(z)\right] dz$ and $B(s)$ is defined in (7.3.1).

Example 7.7. Under the multi-factor CIR model, $C^{RT}(t, \mathbf{x})$ is found again from (7.3.2), where $C^0(t, \mathbf{x})$ is given in (7.3.1), and $\beta(t, \mathbf{x})$ is computed from (7.3.1) with $\mathbf{w}^\lambda = \mathbf{0}$ in (7.3.1) and (7.3.1).

As for the recovery of market value, we assume that at default the recovery is c times the pre-default value $C_{\tau_d-}^{RMV}$. The market pre-default price is given by

$$C^{RMV}(t, \mathbf{X}_t) = \mathbb{E}^{\mathbb{Q}}\{e^{-\int_t^T (r_v + (1-c)\lambda_v)dv} | \mathcal{F}_t\}, \quad 0 \le t \le T.$$

The corresponding drift function can be obtained by applying the quadruple $(1, 0, cC^{RMV}, \tau_d)$ to Theorem 7.5.

Example 7.8. Under the OU model in Section 7.2, the price function $C^{RMV}(t, r, \lambda)$ is given by

$$C^{RMV}(t, r, \lambda) = e^{\hat{A}(T-t)-B(T-t)r-\hat{D}(T-t)\lambda},$$

where $B(s)$ is defined in (7.3.1),
$$\hat{D}(s) = \frac{(1-c)(1-e^{-\kappa_\lambda s})}{\kappa_\lambda}, \quad \text{and}$$
$$\hat{A}(s) = \int_0^s [\frac{1}{2}\sigma_r^2 B^2(z) + \rho\mu\sigma_r\sigma_\lambda B(z)\hat{D}(z) + \frac{1}{2}\mu^2\sigma_\lambda^2\hat{D}^2(z)$$
$$- \kappa_r\theta_r B(z) - \mu\kappa_\lambda\theta_\lambda\hat{D}(z)]\, dz.$$

Example 7.9. Under the multi-factor CIR model, $C^{RMV}(t, \mathbf{x})$ admits the same formula as (7.3.1) but with $\boldsymbol{w}^\lambda$ replaced by $(1-c)\boldsymbol{w}^\lambda$ in (7.3.1) and (7.3.1).

7.3.3   *Optimal Liquidation of CDS*

In this section we consider optimally liquidating a digital CDS position. The investor is a protection buyer who pays a fixed premium to the protection seller from time 0 until default or maturity T, whichever comes first. The premium rate p_0^m, called the market spread, is specified at contract inception. In return, the protection buyer will receive \$1 if default occurs at or before T. The liquidation of the CDS position at time t can be achieved by entering a CDS contract as a protection seller with the same credit reference and same maturity T at the prevailing market spread p_t^m. By definition, the prevailing market spread p_t^m makes the values of two legs equal at time t, i.e.
$$\mathbb{E}^\mathbb{Q}\{\int_t^T e^{-\int_t^u r_v dv} p_t^m \mathbf{1}_{\{u<\tau_d\}} du | \mathcal{G}_t\} = \mathbb{E}^\mathbb{Q}\{e^{-\int_t^{\tau_d} r_v dv} \mathbf{1}_{\{t<\tau_d\leq T\}} | \mathcal{G}_t\} \quad (7.25)$$
If the liquidation occurs at time t, she receives the premium at rate p_t^m and pays the premium at rate p_0^m until default or maturity T. If default occurs, then the default payments from both CDS contracts will cancel. Considering the resulting expected cash flows and (7.25), the mark-to-market value of the CDS is given by
$$\mathbb{E}^\mathbb{Q}\{\int_t^T e^{-\int_t^u r_v dv}(p_t^m - p_0^m)\mathbf{1}_{\{u<\tau_d\}} du | \mathcal{G}_t\}$$
$$= \mathbf{1}_{\{t<\tau_d\}}\mathbb{E}^\mathbb{Q}\{\int_t^T e^{-\int_t^u (r_v+\lambda_v) dv}(\lambda_u - p_0^m) du | \mathcal{F}_t\}$$
$$=: \mathbf{1}_{\{t<\tau_d\}} C^{CDS}(t, \mathbf{X}_t). \quad (7.26)$$

For CDS, we apply the quadruple $(0, -p_0^m, 1, \tau_d)$ to Theorem 7.5 and obtain the drift function:
$$G^{CDS}(t, \mathbf{x}) = -(\nabla_x C^{CDS}(t, \mathbf{x}))^T \Sigma(t, \mathbf{x})\phi^{\tilde{\mathbb{Q}}, \mathbb{Q}}(t, \mathbf{x})$$
$$+ (1 - C^{CDS}(t, \mathbf{x}))(\tilde{\mu}(t, \mathbf{x}) - \mu(t, \mathbf{x}))\hat{\lambda}(t, \mathbf{x}). \quad (7.27)$$

If there is no discrepancy over mark-to-market risk premium, i.e. $\phi^{\tilde{Q},Q}(t,\mathbf{x}) = \mathbf{0}$, then the sign of G^{CDS} is determined by $\tilde{\mu}(t,\mathbf{x}) - \mu(t,\mathbf{x})$ since $C^{CDS} \leq 1$. From this we infer that higher event risk premium (relative to market) implies delayed liquidation.

In general, the market pre-default value C^{CDS} can be solved by PDE (7.13). If the state vector $\mathbf{X}$ admits OU or CIR dynamics, C^{CDS}, and thus G^{CDS}, is given in closed form, as illustrated in the following two examples.

Example 7.10. Under the OU dynamics, the pre-default value of CDS (see (7.26)) is given by the following integral:

$$C^{CDS}(t,r,\lambda) = \int_t^T C^0(t,r,\lambda)\left[\lambda e^{-\kappa_\lambda(u-t)} + \int_t^u e^{-\kappa_r(u-s)}g(s,u)ds - p_0^m\right]du,$$

where $C^0(t,r,\lambda) \equiv C^0(t,r,\lambda;u)$ is given by (7.3.1) with $T = u$ and

$$g(s,u) := \mu\kappa_\lambda\theta_\lambda - \rho\mu\sigma_r\sigma_\lambda\frac{1 - e^{-\kappa_r(u-s)}}{\kappa_r} - (\mu\sigma_\lambda)^2\frac{1 - e^{-\kappa_\lambda(u-s)}}{\kappa_\lambda}.$$

Example 7.11. Under the multi-factor CIR dynamics, the pre-default value $C^{CDS}(t,\mathbf{x})$ of CDS is given by the following integral:

$$\int_t^T C^0(t,\mathbf{x};u)\left[\sum_{i=1}^n\left(\mu w_i^\lambda\left(\kappa_i\theta_i B_i(u-t) + B_i'(u-t)x_i\right)\right) - p_0^m\right]du,$$

where $C^0(t,\mathbf{x};u)$ is given in (7.3.1) with $T = u$ and $B_i(s)$ in (7.3.1).

Example 7.12. For a *forward* CDS with start date $T_a < T$, the protection buyer pays premium at rate p_a from T_a until τ_d or maturity T, and receives 1 if $\tau_d \in [T_a, T]$. By direct computation, the pre-default market value is $C^{CDS}(t,\mathbf{x};T) - C^{CDS}(t,\mathbf{x};T_a)$, $t < T_a$. Consequently, closed-form formulas for the drift function are available under OU or CIR dynamics by Examples 7.10 and 7.11.

We consider a numerical example where interest rate is constant and state vector $\mathbf{X} = \lambda$ follows the CIR dynamics. We assume that the investor agrees with the market on all parameters except the speed of mean reversion for default intensity. In Figure 7.3(a) with $\kappa_\lambda = 0.2 < 0.3 = \tilde{\kappa}_\lambda$, the optimal liquidation strategy is to sell as soon as the market CDS value reaches an upper boundary. In the case with $\kappa_\lambda = 0.3 > 0.2 = \tilde{\kappa}_\lambda$ (see Figure 7.3 (b)), the sell region is below the continuation region.

Remark 7.13. As a straightforward generalization under our framework, one can replace the unit payment at default by $R^{CDS}(\tau_d, \mathbf{X}_{\tau_d})$. Then, we can apply the quadruple $(0, -p_0^m, R^{CDS}, \tau_d)$ to Theorem 7.5, and obtain the same drift function G^{CDS} in (7.27) except with 1 replaced by $R^{CDS}(t,\mathbf{x})$.

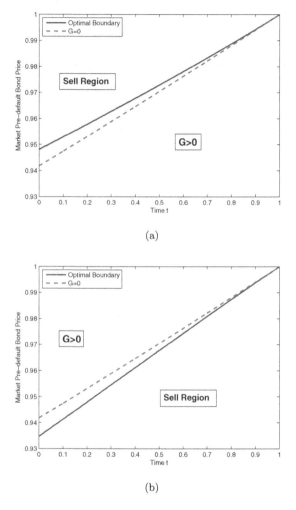

(a)

(b)

Fig. 7.2   Optimal liquidation boundary in terms of market pre-default bond price under CIR dynamics. We take $T = 1$, $r = 0.03$, $\sigma = 0.07$, $\mu = \tilde{\mu} = 2$, and $\theta_\lambda = \tilde{\theta}_\lambda = 0.015$. Panel (a): When $\kappa_\lambda = 0.2 < 0.3 = \tilde{\kappa}_\lambda$, the optimal boundary increases from 0.948 to 1 over time. Panel (b): When $\kappa_\lambda = 0.3 > 0.2 = \tilde{\kappa}_\lambda$, the optimal boundary increases from 0.935 to 1 over time.

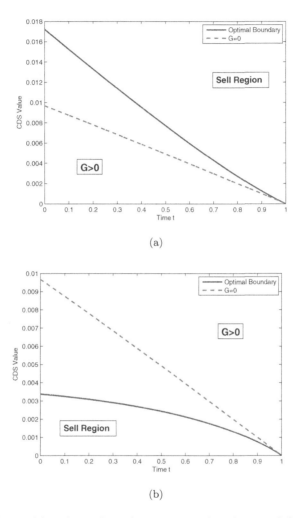

(a)

(b)

Fig. 7.3 Optimal liquidation boundary in terms of market pre-default CDS value under CIR dynamics. We take $T = 1$, $r = 0.03$, $\sigma = 0.07$, $p_0^m = 0.02$, $\mu = \tilde{\mu} = 2$, and $\theta_\lambda = \tilde{\theta}_\lambda = 0.015$. Panel (a): When $\kappa_\lambda = 0.2 < 0.3 = \tilde{\kappa}_\lambda$, liquidation occurs at an upper boundary that decreases from 0.0172 to 0 over $t \in [0, 1]$. Panel (b): When $\kappa_\lambda = 0.3 > 0.2 = \tilde{\kappa}_\lambda$, the CDS is liquidated at a lower liquidation boundary, which decreases from 0.00338 to 0 over time. In both cases, the dashed line defined by $G = 0$ lies within the continuation region.

7.3.4 *Jump-Diffusion Default Intensity*

We can extend our analysis to incorporate jumps to the stochastic state vector. To illustrate this, suppose the default intensity and interest rate are driven by a n-dimensional state vector $\mathbf{X}'$ with the affine jump-diffusion dynamics:

$$d\mathbf{X}'_t = a(t, \mathbf{X}'_t)dt + \Sigma(t, \mathbf{X}'_t)d\mathbf{W}^{\mathbb{P}}_t + d\mathbf{J}_t,$$

where $\mathbf{J} = (J^1, \ldots, J^n)^T$ is a vector of n independent pure jump processes taking values in $\mathbb{R}^n$. Under historical measure $\mathbb{P}$, we assume Markovian jump intensity of the form $\hat{\mathbf{\Lambda}}(t, \mathbf{X}'_t) = (\hat{\Lambda}^1(t, \mathbf{X}'_t), \ldots, \hat{\Lambda}^n(t, \mathbf{X}'_t))^T$ for $\mathbf{J}$. All random jump sizes $(Y^i_j)_{ij}$ of $\mathbf{J}$ are independent, and for each J^i, the associated jump sizes $Y^i_1, Y^i_2, \ldots$ have a common probability density function $\hat{f}^i$.

The default intensity of defaultable security is given by $\hat{\lambda}(t, \mathbf{X}'_t)$ for some positive measurable function $\hat{\lambda}(\cdot, \cdot)$, and the default counting process associated with default time τ_d is denoted by $N_t = \mathbf{1}_{\{t \geq \tau_d\}}$. We denote $(\mathcal{G}_t)_{0 \leq t \leq T}$ to be the full filtration generated by $\mathbf{W}^{\mathbb{P}}, \mathbf{J}$, and τ_d.

We define a market pricing measure $\mathbb{Q}$ in terms of the mark-to-market risk premium $\phi^{\mathbb{Q}, \mathbb{P}}$ and event risk premium μ, which are Markovian and satisfy $\int_0^T \|\phi^{\mathbb{Q}, \mathbb{P}}_u\|^2 du < \infty$ and $\int_0^T \mu_u \hat{\lambda}_u du < \infty$. Due to the presence of $\mathbf{J}$, the market measure $\mathbb{Q}$ can scale the jump intensity of $\mathbf{J}$ by the positive Markovian factors $\delta^i_t = \delta^i(t, \mathbf{X}'_t)$, with $\int_0^T \delta^i_u \hat{\Lambda}^i_u du < \infty$ for $i = 1, \ldots, n$. Also, $\mathbb{Q}$ can transform the jump size distribution of $\mathbf{J}$ by a function $(h^i)_{i=1,\ldots,n} > 0$ satisfying $\int_0^\infty h^i(y) \hat{f}^i(y) dy = 1$ for $i = 1, \ldots, n$.

The Radon-Nikodym derivative is given by

$$\frac{d\mathbb{Q}}{d\mathbb{P}} \Big|\mathcal{G}_t = \mathcal{E}\big(-\phi^{\mathbb{Q},\mathbb{P}} \cdot \mathbf{W}^{\mathbb{P}}\big)_t \mathcal{E}\big((\mu - 1) M^{\mathbb{P}}\big)_t K^{\mathbb{Q},\mathbb{P}}_t,$$

where first two Doléans-Dade exponentials are defined in (7.7) and (7.8) respectively, $M^{\mathbb{P}}_t := N_t - \int_0^t (1 - N_u)\hat{\lambda}_u du$ is the compensated $\mathbb{P}$-martingale associated with N, and the last term

$$K^{\mathbb{Q},\mathbb{P}}_t := \prod_{i=1}^n \Bigg[\exp\Bigg(\int_0^t \int_{\mathbb{R}^n} \big(1 - \delta^i(u, \mathbf{X}'_u) h^i(y)\big) \hat{\Lambda}^i(u, \mathbf{X}'_u) \hat{f}^i(y) dy du \Bigg)$$
$$\prod_{j=1}^{N^{(i)}_t} \big(\delta^i(T^i_j, \mathbf{X}'_{T^i_j}) h^i(Y^i_j)\big) \Bigg],$$

where T^i_j is the jth jump time of J^i and $N^{(i)}_t := \sum_{j \geq 1} \mathbf{1}_{\{T^i_j \leq t\}}$ is the counting process associated with J^i.

By Girsanov Theorem, $\mathbf{W}_t^{\mathbb{Q}} := \mathbf{W}_t^{\mathbb{P}} + \int_0^t \phi_u^{\mathbb{Q},\mathbb{P}} du$ is a $\mathbb{Q}$-Brownian motion, the jump intensity of J^i under $\mathbb{Q}$ is $\Lambda^i(t, \mathbf{X}_t') := \delta^i(t, \mathbf{X}_t')\hat{\Lambda}^i(t, \mathbf{X}_t')$, and the jump size pdf of J^i under $\mathbb{Q}$ is $f^i(y) := h^i(y)\hat{f}^i(y)$. The $\mathbb{Q}$-dynamics of state vector $\mathbf{X}'$ is given by

$$d\mathbf{X}_t' = b(t, \mathbf{X}_t')dt + \Sigma(t, \mathbf{X}_t')d\mathbf{W}_t^{\mathbb{Q}} + d\mathbf{J}_t.$$

Also incorporating the event risk premium, the $\mathbb{Q}$-default intensity is $\lambda(t, \mathbf{X}_t') := \mu(t, \mathbf{X}_t')\hat{\lambda}(t, \mathbf{X}_t')$.

Under the investor's measure $\tilde{\mathbb{Q}}$, we replace b with $\tilde{b}$, $\phi^{\mathbb{Q},\mathbb{P}}$ with $\phi^{\tilde{\mathbb{Q}},\mathbb{P}}$, and $\mathbf{W}^{\mathbb{Q}}$ with $\mathbf{W}^{\tilde{\mathbb{Q}}}$ for the dynamics of $\mathbf{X}'$. For each J^i, the $\tilde{\mathbb{Q}}$-intensity is denoted by $\tilde{\Lambda}^i(t, \mathbf{X}_t') := \tilde{\delta}^i(t, \mathbf{X}_t')\hat{\Lambda}^i(t, \mathbf{X}_t')$, and the jump size pdf under $\tilde{\mathbb{Q}}$ is $\tilde{f}^i(y) := \tilde{h}^i(y)\hat{f}^i(y)$. With investor's event risk premium $\tilde{\mu}$, the default intensity under $\tilde{\mathbb{Q}}$ is $\tilde{\lambda}(t, \mathbf{X}_t') := \tilde{\mu}(t, \mathbf{X}_t')\hat{\lambda}(t, \mathbf{X}_t')$.

The two pricing measures $\mathbb{Q}$ and $\tilde{\mathbb{Q}}$ are related by the Radon-Nikodym derivative:

$$\frac{d\tilde{\mathbb{Q}}}{d\mathbb{Q}}\Big|\mathcal{G}_t = \mathcal{E}\big(-\phi^{\tilde{\mathbb{Q}},\mathbb{Q}}\cdot\mathbf{W}^{\mathbb{Q}}\big)_t \mathcal{E}\big((\frac{\tilde{\mu}}{\mu}-1)M^{\mathbb{Q}}\big)_t K_t^{\tilde{\mathbb{Q}},\mathbb{Q}},$$

where $M_t^{\mathbb{Q}} := N_t - \int_0^t(1 - N_u)\lambda_u du$ is the compensated $\mathbb{Q}$-martingale associated with N, the first two Doléans-Dade exponentials are defined in (7.9) and (7.10), and

$$K_t^{\tilde{\mathbb{Q}},\mathbb{Q}} := \prod_{i=1}^n \Bigg[\exp\bigg(\int_0^t \int_{\mathbb{R}^n} \big(\Lambda^i(u, \mathbf{X}_u')f^i(y) - \tilde{\Lambda}^i(u, \mathbf{X}_u')\tilde{f}^i(y)\big)dydu \bigg)$$
$$\prod_{j=1}^{N_t^{(i)}} \frac{\tilde{\Lambda}^i(T_j^i, \mathbf{X}_{T_j^i}')\tilde{f}^i(\mathbf{Y}_j^i)}{\Lambda^i(T_j^i, \mathbf{X}_{T_j^i}')f^i(\mathbf{Y}_j^i)}\Bigg].$$

Consequently, on top of the mark-to-market risk and event risk premia, the investor can potentially disagree with the market over jump intensity and jump size distribution of $\mathbf{X}'$, allowing for a richer structure of price discrepancy as well as the optimal liquidation strategy.

As in Theorem 7.5, we compute the drift function in terms of pre-default price C and default risk premia, namely,

$$G^J(t, \mathbf{x}) = -\big(\nabla_x C(t, \mathbf{x})\big)^T \Sigma(t, \mathbf{x})\phi^{\tilde{\mathbb{Q}},\mathbb{Q}}(t, \mathbf{x})$$
$$+ (R(t, \mathbf{x}) - C(t, \mathbf{x}))\big(\tilde{\mu}(t, \mathbf{x}) - \mu(t, \mathbf{x})\big)\hat{\lambda}(t, \mathbf{x})$$
$$+ \sum_{i=1}^n \bigg(\int_{\mathbb{R}^n} \big(C(t, \mathbf{x} + y\mathbf{e}_i) - C(t, \mathbf{x})\big)\big(\tilde{\Lambda}^i(t, \mathbf{x})\tilde{f}^i(y) - \Lambda^i(t, \mathbf{x})f^i(y)\big)dy \bigg),$$

where $\mathbf{e}_i := (0, \ldots, 1, \ldots, 0)^T$. We observe that the first two components of G^J share the same functional form as G in (7.16), though the price function C is derived from the jump-diffusion model. Even if the investor and the market assign the same mark-to-market risk and event risk premia, discrepancy over jump intensity and distribution will yield different liquidation strategies. Under quite general affine jump-diffusion models, Duffie *et al.* (2000) provide an analytical treatment of transform analysis, which can be used for the computation of our drift function.

7.4 Optimal Liquidation of Credit Default Index Swaps

We proceed to discuss the optimal liquidation of multi-name credit derivatives. In the literature, there exist many proposed models for modeling multiple defaults and pricing multi-name credit derivatives. Within the intensity-based framework, one popular approach is to model each default time by the first jump of a doubly-stochastic process. The dependence among defaults can be incorporated via some common stochastic factors. This well-known bottom-up valuation framework has been studied in Duffie and Garleanu (2001); Mortensen (2006), among many others.

As a popular alternative, the top-down approach describes directly the dynamics of the cumulative credit portfolio loss, without detailed references to the constituent single names.[4] For our analysis, rather than proposing a new multi-name credit risk model, we adopt the self-exciting top-down model developed by Errais *et al.* (2010). In particular, we will focus on the optimal liquidation of a credit default index swap.

First, we model successive default arrivals by a counting process $(N_t)_{0 \le t \le T}$, and the accumulated portfolio loss by $\Upsilon_t = l_1 + \ldots + l_{N_t}$, with each l_n representing the random loss at the nth default. Under the historical measure $\mathbb{P}$, the default intensity evolves according to the jump-diffusion:

$$d\hat{\lambda}_t = \hat{\kappa}(\hat{\theta} - \hat{\lambda}_t)dt + \sigma\sqrt{\hat{\lambda}_t}\,dW_t^{\mathbb{P}} + \eta\,d\Upsilon_t, \qquad (7.28)$$

where $W^{\mathbb{P}}$ is a standard $\mathbb{P}$-Brownian motion. We assume that the random losses (l_n) are independent with an identical probability density function $\hat{m}$ on $(0, \infty)$. According to the last term in (7.28), each default arrival will increase default intensity $\hat{\lambda}$ by the loss at default scaled by the positive parameter η. This term captures default clustering observed in the multi-name credit derivatives. We assume a constant risk-free interest rate r for

---

[4]Some examples of top-down models include Brigo *et al.* (2007); Ding *et al.* (2009); Longstaff and Rajan (2008); Lopatin and Misirpashaev (2008).

simplicity, and denote $(\mathcal{H}_t)_{0 \leq t \leq T}$ to be the full filtration generated by N, Υ, and $W^{\mathbb{P}}$.

The market measure $\mathbb{Q}$ is characterized by several key components. First, the market's mark-to-market risk premium is assumed to be of the form

$$\phi_t^{\mathbb{Q},\mathbb{P}} = \frac{\hat{\kappa}(\hat{\theta} - \hat{\lambda}_t) - \kappa(\theta - \hat{\lambda}_t)}{\sigma\sqrt{\hat{\lambda}_t}} \tag{7.29}$$

such that the default intensity in (7.28) preserves mean-reverting dynamics with different parameters κ and θ under the market measure $\mathbb{Q}$. Secondly, we assume that the $\mathbb{Q}$-default intensity is $\lambda_t := \mu\hat{\lambda}_t$, with a positive constant event risk premium. Thirdly, the distribution of random losses can be scaled under $\mathbb{Q}$. Specifically, we assume that under $\mathbb{Q}$ the losses (l_n) admit the pdf $m(z) := h(z)\hat{m}(z)$, for some strictly positive function h with $\int_0^\infty h(z)\hat{m}(z)dz = 1$. Then, the Radon-Nikodym derivative associated with $\mathbb{Q}$ and $\mathbb{P}$ is

$$\frac{d\mathbb{Q}}{d\mathbb{P}}\Big|\mathcal{H}_t = \mathcal{E}\big(-\phi^{\mathbb{Q},\mathbb{P}}W^{\mathbb{P}}\big)_t \hat{K}_t^{\mathbb{Q},\mathbb{P}},$$

where $\mathcal{E}\big(-\phi^{\mathbb{Q},\mathbb{P}}W^{\mathbb{P}}\big)$ is defined in (7.7), and

$$\hat{K}_t^{\mathbb{Q},\mathbb{P}} := \exp\left(\int_0^t \int_0^\infty \big(1 - \mu h(z)\big)\hat{\lambda}_u\hat{m}(z)dzdu\right) \prod_{i=1}^{N_t} \big(\mu h(l_i)\big).$$

Under the market pricing measure $\mathbb{Q}$, the $\mathbb{Q}$-default intensity evolves according to:

$$d\lambda_t = \kappa(\mu\theta - \lambda_t)dt + \sigma\sqrt{\mu\lambda_t}\,dW_t^{\mathbb{Q}} + \mu\eta\,d\Upsilon_t,$$

where $W_t^{\mathbb{Q}} := W_t^{\mathbb{P}} + \int_0^t \phi_u^{\mathbb{Q},\mathbb{P}}du$ is a standard $\mathbb{Q}$-Brownian motion. Similarly, we can define the investor's pricing measure $\tilde{\mathbb{Q}}$ through the investor's mark-to-market risk premium $\phi^{\tilde{\mathbb{Q}},\mathbb{P}}$ as in (7.29) with parameters $\tilde{\kappa}$ and $\tilde{\theta}$; default intensity $\tilde{\lambda}_t = \tilde{\mu}\hat{\lambda}_t$ with constant event risk premium $\tilde{\mu}$; and loss scaling function $\tilde{h}$ so that the loss pdf $\tilde{m}(z) = h(z)\hat{m}(z)$.

The credit default index swap is written on a standardized portfolio of H reference entities, such as single-name default swaps, with same notional normalized to 1 and same maturity T. The investor is a protection buyer who pays at the premium rate p_0^m in return for default payments over $(0, T]$. Here, the default payment is assumed to be paid at the time when default occurs, and the premium payment is paid continuously with premium notional equal to $H - N_t$.

The market's cumulative value of the credit default index swap for the protection buyer is equal to the difference between the market values of the default payment leg and premium leg, namely,

$$P_t^{CDX} = \mathbb{E}^{\mathbb{Q}}\Big\{ \int_{(0,T]} e^{-r(u-t)}\, d\Upsilon_u \,|\mathcal{H}_t \Big\}$$

$$- \mathbb{E}^{\mathbb{Q}}\Big\{ p_0^m \int_{(0,T]} e^{-r(u-t)}(H - N_u)\, du \,|\mathcal{H}_t \Big\}, \quad t \leq T. \quad (7.30)$$

Hence, similar to (7.3), the protection buyer solves the following optimal stopping problem:

$$V_t^{CDX} = \operatorname*{ess\,sup}_{\tau \in \mathcal{T}_{t,T}} \mathbb{E}^{\tilde{\mathbb{Q}}}\{ e^{-r(\tau-t)} P_\tau^{CDX} \,|\mathcal{H}_t \}. \quad (7.31)$$

The associated delayed liquidation premium is defined by

$$L_t^{CDX} = V_t^{CDX} - P_t^{CDX}. \quad (7.32)$$

The derivation of the optimal liquidation strategy involves computing the market's ex-dividend value, defined by

$$C_t^{CDX} = \mathbb{E}^{\mathbb{Q}}\Big\{ \int_{(t,T]} e^{-r(u-t)}\, d\Upsilon_u \,|\mathcal{H}_t \Big\}$$

$$- \mathbb{E}^{\mathbb{Q}}\Big\{ p_0^m \int_{(t,T]} e^{-r(u-t)}(H - N_u)\, du \,|\mathcal{H}_t \Big\}. \quad (7.33)$$

Proposition 7.14. *The market's ex-dividend value of the credit default index swap in (7.33) can be expressed as* $C_t^{CDX} = C^{CDX}(t, \lambda_t, N_t)$, *where*

$$C^{CDX}(t, \lambda, n) = k_2(t, T)\lambda + k_1(t, T)n + k_0(t, T), \quad (7.34)$$

for $t \leq T$, with coefficients

$$k_2(t, T) = (cr + p_0^m)\Big(\frac{e^{-(\rho+r)(T-t)}}{\rho(\rho+r)} - \frac{e^{-r(T-t)}}{\rho r} + \frac{1}{r(\rho+r)} \Big)$$

$$+ \frac{ce^{-r(T-t)}}{\rho}\big(1 - e^{-\rho(T-t)}\big),$$

$$k_1(t, T) = \frac{p_0^m\big(1 - e^{-r(T-t)}\big)}{r},$$

$$k_0(t, T) = \Big((rc + p_0^m)\big[e^{-r(T-t)}\big(\frac{1}{r\rho} - \frac{e^{-\rho(T-t)}}{\rho(\rho+r)} - \frac{T-t}{r} - \frac{1}{r^2} \big) + \frac{\rho}{r^2(r+\rho)} \big]$$

$$+ ce^{-r(T-t)}\big(\frac{e^{-\rho(T-t)} - 1}{\rho} + T - t \big) \Big) \frac{\kappa\mu\theta}{\rho} - \frac{p_0^m H}{r}\big(1 - e^{-r(T-t)}\big),$$

and constants

$$c = \int_0^\infty zm(z)\,dz, \quad \text{and} \quad \rho = \kappa - \mu\eta c. \quad (7.35)$$

Proof. Using integration by parts, we re-write the market's ex-dividend value as

$$C_t^{CDX} = e^{-r(T-t)} \mathbb{E}^{\mathbb{Q}}\{\Upsilon_T | \mathcal{H}_t\} - \Upsilon_t + \int_t^T e^{-r(u-t)} \big[r\mathbb{E}^{\mathbb{Q}}\{\Upsilon_u | \mathcal{H}_t\}$$
$$- p_0^m \big(H - \mathbb{E}^{\mathbb{Q}}\{N_u | \mathcal{H}_t\} \big) \big] \, du. \tag{7.36}$$

Hence, the computation of C^{CDX} involves calculating $\mathbb{E}^{\mathbb{Q}}\{N_u \mid \mathcal{H}_t\}$ and $\mathbb{E}^{\mathbb{Q}}\{\Upsilon_u \mid \mathcal{H}_t\}$, $u \geq t$. Since default intensity λ follows a square-root jump-diffusion dynamics, these conditional expectation admit the closed-form expressions:

$$\mathbb{E}^{\mathbb{Q}}\{N_u \mid \lambda_t = \lambda, N_t = n, \Upsilon_t = v\} = \mathcal{A}(t,u) + \mathcal{B}(t,u)\lambda + n, \tag{7.37}$$
$$\mathbb{E}^{\mathbb{Q}}\{\Upsilon_u \mid \lambda_t = \lambda, N_t = n, \Upsilon_t = v\} = c\mathcal{A}(t,u) + c\mathcal{B}(t,u)\lambda + v, \tag{7.38}$$

for $t \leq u \leq T$, where

$$\mathcal{A}(t,u) = \frac{\kappa\mu\theta}{\kappa - \mu\eta c} \Big(\frac{e^{-(\kappa - \mu\eta c)(u-t)} - 1}{\kappa - \mu\eta c} + u - t \Big),$$

$$\mathcal{B}(t,u) = \frac{1}{\kappa - \mu\eta c}(1 - e^{-(\kappa - \mu\eta c)(u-t)}).$$

Here, c is the market's expected loss at default given in (7.35). Substituting (7.37) and (7.38) into (7.36), we obtain the closed-form formula for market's ex-dividend value in (7.34). $\qquad\square$

As a result, the ex-dividend value C^{CDX} is linear in the default intensity λ_t and number of defaults N_t. Next, we characterize the optimal corresponding liquidation premium and strategy.

Theorem 7.15. *Under the top-down credit risk model in (7.28), the delayed liquidation premium associated with the credit default index swap is given by*

$$L^{CDX}(t,\lambda) = \sup_{\tau \in \mathcal{T}_{t,T}} \mathbb{E}^{\tilde{\mathbb{Q}}}\{ \int_t^\tau e^{-r(u-t)} G^{CDX}(u, \lambda_u) du \mid \lambda_t = \lambda \}, \tag{7.39}$$

where

$$G^{CDX}(t,\lambda) = k_2(t,T)\mu(\tilde{\kappa}\tilde{\theta} - \kappa\theta) \tag{7.40}$$
$$+ \Big((\mu\eta k_2(t,T) + 1)(\frac{\tilde{\mu}\tilde{c}}{\mu} - c) + k_1(t,T)(\frac{\tilde{\mu}}{\mu} - 1) - k_2(t,T)(\tilde{\kappa} - \kappa) \Big)\lambda,$$

with $\tilde{c} := \int_0^\infty z\tilde{m}(z)dz$. *If* $G^{CDX}(t,\lambda) \geq 0 \; \forall(t,\lambda)$, *then it is optimal to delay the liquidation till maturity* T. *If* $G^{CDX}(t,\lambda) \leq 0 \; \forall(t,\lambda)$, *then it is optimal to sell immediately.*

Proof. In view of the definition of L^{CDX} in (7.32), we consider the dynamics of P^{CDX}. First, it follows from (7.30) and (7.33) that

$$e^{-r(u-t)}P_u^{CDX} = e^{-r(u-t)}C_u^{CDX} + \int_{(0,u]} e^{-r(v-t)}\left(d\Upsilon_v - p_0^m(H - N_v)dv\right).$$

$$(7.41)$$

Using (7.41) and the fact that $e^{-rt}P_t^{CDX}$ is $\mathbb{Q}$-martingale (whose SDE must have no drift), we apply Ito's lemma to get

$$e^{-r(\tau-t)}P_\tau^{CDX} - P_t^{CDX}$$

$$= \int_t^\tau e^{-r(u-t)}\frac{\partial C^{CDX}}{\partial\lambda}(u,\lambda_u,N_u)\sigma\sqrt{\mu\lambda_u}dW_u^{\mathbb{Q}}$$

$$+ \left[\sum_{t<u\leq\tau} e^{-r(u-t)}(\Upsilon_u - \Upsilon_{u-}) - \int_t^\tau\int_0^\infty e^{-r(u-t)}zm(z)\lambda_u\,dz\,du\right]$$

$$+ \left[\sum_{t<u\leq\tau} e^{-r(u-t)}\left(C^{CDX}(u,\lambda_u,N_u) - C^{CDX}(u,\lambda_{u-},N_{u-})\right)\right.$$

$$- \int_t^\tau\int_0^\infty e^{-r(u-t)}\left(C^{CDX}(u,\lambda_u + \mu\eta z, N_u + 1)\right.$$

$$\left.- C^{CDX}(u,\lambda_u,N_u)\right)m(z)\lambda_u\,dz\,du\right] \qquad (7.42)$$

$$= \int_t^\tau e^{-r(u-t)}\left(\frac{\partial C^{CDX}}{\partial\lambda}(u,\lambda_u,N_u)\sigma\sqrt{\mu\lambda_u}dW_u^{\tilde{\mathbb{Q}}} + G^{CDX}(u,\lambda_u,N_u)du\right)$$

$$+ \left[\sum_{t<u\leq\tau} e^{-r(u-t)}(\Upsilon_u - \Upsilon_{u-}) - \int_t^\tau\int_0^\infty e^{-r(u-t)}z\tilde{m}(z)\tilde{\lambda}_u\,dz\,du\right]$$

$$+ \left[\sum_{t<u\leq\tau} e^{-r(u-t)}\left(C^{CDX}(u,\lambda_u,N_u) - C^{CDX}(u,\lambda_{u-},N_{u-})\right)\right.$$

$$- \int_t^\tau\int_0^\infty e^{-r(u-t)}\left(C^{CDX}(u,\lambda_u + \mu\eta z, N_u + 1)\right.$$

$$\left.- C^{CDX}(u,\lambda_u,N_u)\right)\tilde{m}(z)\tilde{\lambda}_u\,dz\,du\right], \qquad (7.43)$$

for $t \leq \tau \leq T$, where

$$G^{CDX}(t,\lambda,n) := \frac{\partial C^{CDX}}{\partial\lambda}(t,\lambda,n)\left((\tilde{\kappa}\tilde{\theta} - \kappa\theta)\mu - (\tilde{\kappa} - \kappa)\lambda\right)$$

$$+ \int_0^\infty (z + C^{CDX}(t,\lambda + \mu\eta z, n + 1) - C^{CDX}(t,\lambda,n))(\frac{\tilde{\mu}}{\mu}\tilde{m}(z) - m(z))\lambda dz.$$

$$(7.44)$$

Note that the two compensated $\mathbb{Q}$-martingale terms in (7.42) account for, respectively, losses and changes in C^{CDX} value due to default arrivals. The second equation (7.43) follows from change of measure from $\mathbb{Q}$ to $\tilde{\mathbb{Q}}$.

By Proposition 7.14, the terms $\frac{\partial C^{CDX}}{\partial \lambda}$ and $C^{CDX}(t, \lambda + \mu \eta z, n + 1) - C^{CDX}(t, \lambda, n)$ do *not* depend on n. Consequently, G^{CDX} does not depend on n, and admits the closed-form formula (7.40) upon a substitution of (7.34) into (7.44).

By taking the expectation on both sides of (7.43) under $\tilde{\mathbb{Q}}$, the delayed liquidation premium L^{CDX} satisfies (7.39) and depends only on t and λ. If $G^{CDX} \geq 0$, then the integrand in (7.39) is positive a.s. and therefore the largest possible stopping time T is optimal. If $G^{CDX} \leq 0$, then $\tau^* = t$ is optimal and $L_t^{CDX} = 0$ a.s. $\qquad\square$

We observe that the drift function consists of two components. The first component in (7.44) accounts for the disagreement between investor and market on the fluctuation of market ex-dividend value, while the second integral term reflects the disagreement on the jumps of market's cumulative value arising from the losses at default and the jumps in the ex-dividend value. Even though the market's cumulative value P^{CDX} in (7.30) and the optimal expected liquidation value V^{CDX} in (7.31) are path-dependent, both the delayed liquidation premium L^{CDX} in (7.39) and G^{CDX} in (7.40) depend only on t and λ due to the special structure of C^{CDX} given in (7.34).

To obtain the variational inequality of L^{CDX}, we recall that $\tilde{\lambda} = \tilde{\mu}\lambda/\mu$ and the $\tilde{\mathbb{Q}}$-dynamics of default intensity λ:

$$d\lambda_t = \tilde{\kappa}(\mu\tilde{\theta} - \lambda_t)dt + \sigma\sqrt{\mu\lambda_t}\,dW_t^{\tilde{\mathbb{Q}}} + \mu\eta\,d\Upsilon_t.$$

The delayed liquidation premium $L^{CDX}(t, \lambda)$ as a function of time t and $\mathbb{Q}$-default intensity λ satisfies the variational inequality

$$\min\left(-\frac{\partial L^{CDX}}{\partial t} - \tilde{\kappa}(\mu\tilde{\theta} - \lambda)\frac{\partial L^{CDX}}{\partial \lambda} - \frac{\sigma^2\mu\lambda}{2}\frac{\partial L^{CDX}}{\partial \lambda^2} + rL^{CDX} \right.$$
$$\left. - \frac{\tilde{\mu}\lambda}{\mu}\int_0^\infty \left(L^{CDX}(t, \lambda + \mu\eta z) - L^{CDX} \right)\tilde{m}(z)dz - G^{CDX}, \; L^{CDX} \right) = 0,$$

$$(7.45)$$

for $(t, \lambda) \in [0, T) \times \mathbb{R}$, with terminal condition $L(T, \lambda) = 0$ for $\lambda \in \mathbb{R}$.

We consider a numerical example for an index swap with constant losses at default. In this case, the integral term in (7.45) reduces to $L^{CDX}(t, \lambda + \mu\eta c) - L^{CDX}(t, \lambda)$, where c is the constant loss. We employ the standard implicit PSOR iterative algorithm to solve L^{CDX} by finite difference method with Neumann condition applied on the intensity boundary. There exist many alternative numerical methods to solve variational inequality with an integral term.[5] We opt to apply a second-order Taylor approximation to the difference $L^{CDX}(t, \lambda + \mu\eta c) - L^{CDX}(t, \lambda) \approx \partial_\lambda L^{CDX}(t, \lambda)\mu\eta c + \frac{1}{2}\partial_{\lambda\lambda} L^{CDX}(t, \lambda)(\mu\eta c)^2$. In turn, these new partial derivatives are incorporated in the existing partial derivatives in (7.45), rendering the variational inequality completely linear in λ, and thus, allowing for rapid computation.

We denote the investor's sell region $\mathcal{S}$ and delay region $\mathcal{D}$ by

$$\mathcal{S}^{CDX} = \{(t, \lambda) \in [0, T] \times \mathbb{R} : L^{CDX}(t, \lambda) = 0\}, \qquad (7.46)$$
$$\mathcal{D}^{CDX} = \{(t, \lambda) \in [0, T] \times \mathbb{R} : L^{CDX}(t, \lambda) > 0\}.$$

On the other hand, we observe from (7.34) a one-to-one correspondence between the market's ex-dividend value C^{CDX} of an index swap and its default intensity λ for any fixed $t < T$, namely,

$$\lambda = \frac{C^{CDX} - k_1(t, T)n - k_0(t, T)}{k_2(t, T)}. \qquad (7.47)$$

Substituting (7.47) into (7.46) and (7.4), we can describe the sell region and delay region in terms of the observable market ex-dividend value C^{CDX}.

In Figure 7.4, we assume that the investor agrees with the market on all parameters except the speed of mean reversion for default intensity. In the case with $\kappa = 0.5 < 1 = \tilde{\kappa}$ (Figure 7.4(a)), the investor's optimal liquidation strategy is to sell as soon as the market ex-dividend value of index swap C^{CDX} reaches an upper boundary. In the case with $\kappa = 1 > 0.5 = \tilde{\kappa}$ (Figure 7.4(b)), the sell region is below the continuation region.

In summary, we have analyzed the optimal liquidation of a credit default index swap under a top-down credit risk model. The selected model and contract specification give us tractable analytical results that are amenable for numerical computation. The top-down model implies that the underlying credit portfolio can experience countably many defaults. This feature is innocuous for a large diversified portfolio in practice since the likelihood of total default is negligible.

---

[5]See, among others, Andersen and Andreasen (2000); d'Halluin *et al.* (2005)

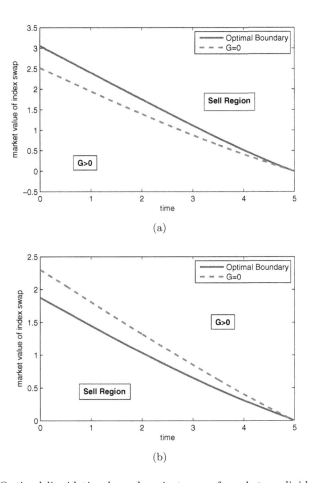

Fig. 7.4 Optimal liquidation boundary in terms of market ex-dividend value of an index swap. We take $T = 5$, $r = 0.03$, $H = 10$, $\eta = 0.25$, $\sigma = 0.5$, $c = \tilde{c} = 0.5$, $p_0^m = 0.02$, $\theta = \tilde{\theta} = 1$, and $\mu = \tilde{\mu} = 1.1$. Panel (a): When $\kappa = 0.5 < 1 = \tilde{\kappa}$, liquidation occurs at an upper boundary that decreases from 3 to 0 over $t \in [0, 5]$. Panel (b): When $\kappa = 1 > 0.5 = \tilde{\kappa}$, the index swap is liquidated at a lower liquidation boundary, which decreases from 1.9 to 0 over time. In both cases, the dashed line defined by $G^{CDX} = 0$ lies within the continuation region.

Our analysis here can be extended to the liquidation of a collateralized debt obligation (CDO) contract. Consider a tranche with lower and higher attachment points $K_1, K_2 \in [0, 1]$ of a CDO with H names, each with a unit notional. The tranche loss is a function of the accumulated loss Υ_t, given by $\tilde{L}_t = (\Upsilon_t - K_1 H)^+ - (\Upsilon_t - K_2 H)^+$, $t \in [0, T]$. With premium rate p_0^m, the ex-dividend market price of the CDO tranche for the protection buyer is

$$
\mathcal{C}^{CDO}(t, \lambda_t, \Upsilon_t) = \mathbb{E}^{\mathbb{Q}} \Big\{ \int_{(t,T]} e^{-r(u-t)} \, d\tilde{L}_u \, | \mathcal{H}_t \Big\}
$$
$$
- \mathbb{E}^{\mathbb{Q}} \Big\{ p_0^m \int_{(t,T]} e^{-r(u-t)} (H(K_2 - K_1) - \tilde{L}_u) \, du \, | \mathcal{H}_t \Big\}.
$$

Hence, the CDO price is a function of the accumulated loss Υ, as opposed to N in the case of CDX (see Proposition 7.14 above).

7.5 Optimal Buying and Selling

Next, we adapt our model to study the optimal buying and selling problem. Consider an investor whose objective is to maximize the revenue through a buy/sell transaction of a defaultable claim (Y, A, R, τ_d) with market price process P in (7.1.1). The problem is studied separately under two scenarios, namely, when the short sale of the defaultable claim is permitted or prohibited. We shall analyze these problems under the Markovian credit risk model in Section 7.2.

If the investor seeks to purchase a defaultable claim from the market, the optimal purchase timing problem and the associated *delayed purchase premium* can be defined as:

$$
V_t^b = \operatorname*{ess\,inf}_{\tau^b \in \mathcal{T}_{t,T}} \mathbb{E}^{\tilde{\mathbb{Q}}} \{ e^{-\int_t^{\tau^b} r_v dv} P_{\tau^b} | \mathcal{G}_t \}, \quad \text{and} \quad L_t^b := P_t - V_t^b \geq 0.
$$

7.5.1 *Optimal Timing with Short Sale Possibility*

When short sale is permitted, there is no restriction on the ordering of purchase time τ^b and sale time τ^s. The investor's investment timing is found from the optimal double-stopping problem:

$$
\mathcal{U}_t := \operatorname*{ess\,sup}_{\tau^b \in \mathcal{T}_{t,T}, \tau^s \in \mathcal{T}_{t,T}} \mathbb{E}^{\tilde{\mathbb{Q}}} \{ e^{-\int_t^{\tau^s} r_v dv} P_{\tau^s} - e^{-\int_t^{\tau^b} r_v dv} P_{\tau^b} | \mathcal{G}_t \}.
$$

Since the defaultable claim will mature at T, we interpret the choice of $\tau^b = T$ or $\tau^s = T$ as no buy/sell transaction at T.

In fact, we can separate $\mathcal{U}$ into two optimal (single) stopping problems. Precisely, we have

$$\mathcal{U}_t = \left(\underset{\tau^s \in \mathcal{T}_{t,T}}{\mathrm{ess\,sup}}\ \mathbb{E}^{\tilde{\mathbb{Q}}}\{e^{-\int_t^{\tau^s} r_v dv} P_{\tau^s} | \mathcal{G}_t\} - P_t \right)$$

$$+ \left(P_t - \underset{\tau^b \in \mathcal{T}_{t,T}}{\mathrm{ess\,inf}}\ \mathbb{E}^{\tilde{\mathbb{Q}}}\{e^{-\int_t^{\tau^b} r_v dv} P_{\tau^b} | \mathcal{G}_t\} \right)$$

$$= L_t + L_t^b.$$

Hence, we have separated $\mathcal{U}$ into a sum of the *delayed liquidation premium* and the *delayed purchase premium*. As a result, the optimal sale time τ^{s*} does not depend on the choice of the optimal purchase time τ^{b*}.

The timing decision again depends crucially on the sub/super-martingale properties of discounted market price under measure $\tilde{\mathbb{Q}}$. Under the Markovian credit risk model in Section 3, we can apply Theorem 7.5 to describe the optimal purchase and sale strategies in terms of the drift function $G(t, \mathbf{x})$ in (7.16).

Proposition 7.16. *If $G(t, \mathbf{x}) \geq 0 \ \forall(t, \mathbf{x}) \in [0, T] \times \mathbb{R}^n$, then it is optimal to immediately buy the defaultable claim and hold it till maturity T, i.e. $\tau^{b*} = t$ and $\tau^{s*} = T$ are optimal for $\mathcal{U}_t$. If $G(t, \mathbf{x}) \leq 0 \ \forall(t, \mathbf{x}) \in [0, T] \times \mathbb{R}^n$, then it is optimal to immediately short sell the claim and maintain the position till T, i.e. $\tau^{s*} = t$ and $\tau^{b*} = T$ are optimal for $\mathcal{U}_t$.*

7.5.2 *Sequential Buying and Selling*

Prohibiting the short sale of defaultable claims implies the ordering: $\tau^b \leq \tau^s \leq T$. Therefore, the investor's value function is

$$U_t := \underset{\tau^b \in \mathcal{T}_{t,T}, \tau^s \in \mathcal{T}_{\tau^b,T}}{\mathrm{ess\,sup}}\ \mathbb{E}^{\tilde{\mathbb{Q}}}\{e^{-\int_t^{\tau^s} r_v dv} P_{\tau^s} - e^{-\int_t^{\tau^b} r_v dv} P_{\tau^b} | \mathcal{G}_t\}. \quad (7.48)$$

The difference $\mathcal{U}_t - U_t \geq 0$ can be viewed as the cost of the short sale constraint to the investor.

As in Section 7.2, we adopt the Markovian credit risk model, and derive from the $\tilde{\mathbb{Q}}$-dynamics of discounted market price in (7.17) to obtain

$$U_t = \underset{\tau^b \in \mathcal{T}_{t,T}, \tau^s \in \mathcal{T}_{\tau^b,T}}{\mathrm{ess\,sup}}\ \mathbb{E}^{\tilde{\mathbb{Q}}}\left\{ \int_{\tau^b}^{\tau^s} (1 - N_u) e^{-\int_t^u r_v dv} G(u, \mathbf{X}_u) du | \mathcal{G}_t \right\}$$

$$= \mathbf{1}_{\{t < \tau_d\}} \underset{\tau^b \in \mathcal{T}_{t,T}, \tau^s \in \mathcal{T}_{\tau^b,T}}{\mathrm{ess\,sup}}\ \mathbb{E}^{\tilde{\mathbb{Q}}}\left\{ \int_{\tau^b}^{\tau^s} e^{-\int_t^u (r_v + \tilde{\lambda}_v) dv} G(u, \mathbf{X}_u) du | \mathcal{F}_t \right\}.$$

Using this probabilistic representation, we immediately deduce the optimal buy/sell strategy in the extreme cases analogues to Theorem 7.5.

Proposition 7.17. *If $G(t, \mathbf{x}) \geq 0 \ \forall (t, \mathbf{x}) \in [0, T] \times \mathbb{R}^n$, then it is optimal to purchase the defaultable claim immediately and hold until maturity, i.e. $\tau^{b*} = t$ and $\tau^{s*} = T$ are optimal for U_t.*

If $G(t, \mathbf{x}) \leq 0 \ \forall (t, \mathbf{x}) \in [0, T] \times \mathbb{R}^n$, then it is optimal to never purchase the claim, i.e. $\tau^{b*} = \tau^{s*} = T$ is optimal for U_t.

Define $\hat{U}(t, \mathbf{X}_t)$ as the pre-default value of U_t, satisfying $U_t := \mathbf{1}_{\{t < \tau_d\}} \hat{U}(t, \mathbf{X}_t)$. We may view $\hat{U}(t, \mathbf{X}_t)$ as a *sequential* optimal stopping problem.

Proposition 7.18. *The value function U_t in (7.48) can be expressed in terms of the delayed liquidation premium $\hat{L}$ in (7.18). Precisely, we have*

$$\hat{U}(t, \mathbf{X}_t) = \operatorname*{ess\,sup}_{\tau^b \in \mathcal{T}_{t,T}} \mathbb{E}^{\tilde{Q}} \left\{ e^{-\int_t^{\tau^b} (r_u + \tilde{\lambda}_u) du} \hat{L}_{\tau^b} \big| \mathcal{F}_t \right\}. \tag{7.49}$$

Proof. We note that, after any purchase time τ^b, the investor will face the liquidation problem V_{τ^b} in (7.3). Then using repeated conditioning, U_t in (7.48) satisfies

$$U_t =$$

$$\operatorname*{ess\,sup}_{\tau^b \in \mathcal{T}_{t,T}, \tau^s \in \mathcal{T}_{\tau^b,T}} \mathbb{E}^{\tilde{Q}} \left\{ \left(e^{-\int_t^{\tau^b} r_u du} \mathbb{E}^{\tilde{Q}} \left\{ e^{-\int_{\tau^b}^{\tau^s} r_u du} P_{\tau^s} \big| \mathcal{G}_{\tau^b} \right\} - e^{-\int_t^{\tau^b} r_u du} P_{\tau^b} \right) \big| \mathcal{G}_t \right\}$$

$$\tag{7.50}$$

$$\leq \operatorname*{ess\,sup}_{\tau^b \in \mathcal{T}_{t,T}} \mathbb{E}^{\tilde{Q}} \left\{ e^{-\int_t^{\tau^b} r_u du} (V_{\tau^b} - P_{\tau^b}) \big| \mathcal{G}_t \right\} \tag{7.51}$$

$$= \operatorname*{ess\,sup}_{\tau^b \in \mathcal{T}_{t,T}} \mathbb{E}^{\tilde{Q}} \left\{ e^{-\int_t^{\tau^b} r_u du} L_{\tau^b} \big| \mathcal{G}_t \right\}$$

$$= \mathbf{1}_{\{t < \tau_d\}} \operatorname*{ess\,sup}_{\tau^b \in \mathcal{T}_{t,T}} \mathbb{E}^{\tilde{Q}} \left\{ e^{-\int_t^{\tau^b} (r_u + \tilde{\lambda}_u) du} \hat{L}_{\tau^b} \big| \mathcal{F}_t \right\}. \tag{7.52}$$

On the other hand, on the RHS of (7.51) we see that $V_{\tau^b} = \mathbb{E}^{\tilde{Q}} \left\{ e^{-\int_{\tau^b}^{\tau^{s*}} r_u du} P_{\tau^{s*}} \big| \mathcal{G}_{\tau^b} \right\}$, with the optimal stopping time $\tau^{s*} := \inf\{ t \geq \tau^b : V_t = P_t \}$ (see (7.6)). This is equivalent to taking the admissible stopping time τ^{s*} for U_t in (7.48), so the reverse of inequality (7.51) also holds. Finally, equating (7.50) and (7.52) and removing the default indicator, we arrive at (7.49). $\qquad \square$

According to Proposition 7.18, the investor, who anticipates to liquidate the defaultable claim after purchase, seeks to maximize the delayed liquidation premium when deciding to buy the derivative from the market. The practical implication of representation (7.49) is that we first solve for the pre-default delayed liquidation premium $\hat{L}(t, \mathbf{x})$ by variational inequality (7.19). Then, using $\hat{L}(t, \mathbf{x})$ as input, we solve $\hat{U}$ by

$$\min\left(-\frac{\partial \hat{U}}{\partial t}(t, \mathbf{x}) - \mathcal{L}_{\tilde{b}, \tilde{\lambda}}\hat{U}(t, \mathbf{x}), \ \hat{U}(t, \mathbf{x}) - \hat{L}(t, \mathbf{x}) \right) = 0, \qquad (7.53)$$

where $\mathcal{L}_{\tilde{b}, \tilde{\lambda}}$ is defined in (7.14), and the terminal condition is $\hat{U}(T, \mathbf{x}) = 0$, for $\mathbf{x} \in \mathbb{R}^n$. In other words, the solution for $\hat{L}(t, \mathbf{x})$ provides the investor's optimal liquidation boundary after the purchase, and the variational inequality for $U(t, \mathbf{x})$ in (7.53) gives the investor's optimal purchase boundary.

In Figure 7.5, we show a numerical example for a defaultable zero-coupon zero-recovery bond where interest rate is constant and λ follows the CIR dynamics. The investor agrees with the market on all parameters except the speed of mean reversion for default intensity. When $\kappa_\lambda < \tilde{\kappa}_\lambda$, the optimal strategy is to buy as soon as the price enters the purchase region and subsequently sell at the (higher) optimal liquidation boundary. When $\kappa_\lambda > \tilde{\kappa}_\lambda$, the optimal liquidation boundary is below the purchase boundary. However, it is possible that the investor buys at a lower price and subsequently sells at a higher price since both boundaries are increasing. It is also possible to buy-high-sell-low, realizing a loss on these sample paths. On average, the optimal sequential buying and selling strategy enables the investor to profit from the price discrepancy. Finally, when short sale is allowed, the investor's strategy follows the corresponding boundaries without the buy-first/sell-later constraint.

7.6   Concluding Remarks

In summary, we have provided a flexible mathematical model for the optimal liquidation of various credit derivatives under price discrepancy. We have identified the situations where the optimal timing is trivial and also solved for the cases when sophisticated strategies are involved. The optimal liquidation framework enables investors to quantify their views on default risk, extract profit from price discrepancy, and perform more effective risk management. Our model can also be modified and extended to incorporate single or multiple buying and selling decisions.

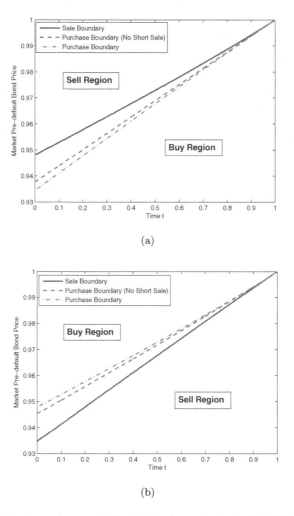

Fig. 7.5 Optimal purchase and liquidation boundaries in the CIR model. The common parameters are $T = 1$, $r = 0.03$, $\sigma = 0.07$, $\mu = \tilde{\mu} = 2$, and $\theta_\lambda = \tilde{\theta}_\lambda = 0.015$. Panel (a): When $\kappa_\lambda = 0.2 < 0.3 = \tilde{\kappa}_\lambda$, the short sale constraint moves the purchase boundary higher. Both purchase boundaries, with or without short sale, are dominated by the liquidation boundary. Panel (b): When $\kappa_\lambda = 0.3 > 0.2 = \tilde{\kappa}_\lambda$, the short sale constraint moves the purchase boundary lower. The liquidation boundary lies below both purchase boundaries.

For future research, a natural direction is to consider credit derivatives trading under other default risk models. For multi-name credit derivatives, in contrast to the top-down approach taken in Section 5, one can consider the optimal liquidation problem under the bottom-up framework. Liquidation problems are also important for derivatives portfolios in general. To this end, the structure of dependency between multiple risk factors is crucial in modeling price dynamics. Moreover, it is both practically and mathematically interesting to allow for early partial/full cancellation for credit derivatives, as is common for mortgages. On the other hand, market participants' pricing rules may vary due to different risk preferences. This leads to the interesting question of how risk aversion influences their derivatives purchase/liquidation timing.

.

Bibliography

Abramowitz, M. and Stegun, I. (1965). *Handbook of Mathematical Functions: with Formulas, Graphs, and Mathematical Tables*, Vol. 55 (Dover Publications).

Alili, L., Patie, P., and Pedersen, J. (2005). Representations of the first hitting time density of an Ornstein-Uhlenbeck process, *Stochastic Models* **21**, 4, pp. 967–980.

Almgren, R. F. (2003). Optimal execution with nonlinear impact functions and trading-enhanced risk, *Applied Mathematical Finance* **10**, pp. 1–18.

Alvarez, L. H. (2003). On the properties of r-excessive mappings for a class of diffusions, *Annals of Applied Probability* , pp. 1517–1533.

Andersen, L. and Andreasen, J. (2000). Jump-diffusion processes: Volatility smile fitting and numerical methods for option pricing, *Review of Derivatives Research* **4**, pp. 231–262.

Anthony, M. and MacDonald, R. (1998). On the mean-reverting properties of target zone exchange rates: Some evidence from the ERM, *European Economic Review* **42**, 8, pp. 1493–1523.

Artzner, P., Delbaen, F., Eber, J., and Heath, D. (1999). Coherent measures of risk, *Mathematical Finance* **9**, 3, pp. 203–228.

Avellaneda, M. and Lee, J.-H. (2010). Statistical arbitrage in the us equities market, *Quantitative Finance* **10**, 7, pp. 761–782.

Azizpour, S., Giesecke, K., and Kim, B. (2011). Premia for correlated default risk, *Journal of Economic Dynamics and Control* **35(8)**, pp. 1340–1357.

Balvers, R., Wu, Y., and Gilliland, E. (2000). Mean reversion across national stock markets and parametric contrarian investment strategies, *The Journal of Finance* **55**, 2, pp. 745–772.

Bensoussan, A. and Lions, J.-L. (1978). *Applications des Inequations Variationalles en Controle Stochstique* (Dunod, Paris).

Bensoussan, A. and Lions, J.-L. (1982). *Applications of variational inequalities in stochastic control* (North-Holland Publishing Co., Amsterdam).

Benth, F. E. and Karlsen, K. H. (2005). A note on Merton's portfolio selection problem for the Schwartz mean-reversion model, *Stochastic Analysis and Applications* **23**, 4, pp. 687–704.

Berndt, A., Douglas, R., Duffie, D., Ferguson, M., and Schranz, D. (2005). Measuring default risk premia from default swap rates and EDFs, Working Paper, Carnegie Mellon University.

Bertram, W. (2010). Analytic solutions for optimal statistical arbitrage trading, *Physica A: Statistical Mechanics and its Applications* **389**, 11, pp. 2234–2243.

Bessembinder, H., Coughenour, J. F., Seguin, P. J., and Smoller, M. M. (1995). Mean reversion in equilibrium asset prices: Evidence from the futures term structure, *The Journal of Finance* **50**, 1, pp. 361–375.

Bielecki, T. R., Jeanblanc, M., and Rutkowski, M. (2008). Pricing and trading credit default swaps in a hazard process model, *Annals of Applied Probability* **18**, 6, pp. 2495–2529.

Bielecki, T. R. and Rutkowski, M. (2002). *Credit Risk: Modeling, Valuation and Hedging* (Springer Finance).

Borodin, A. and Salminen, P. (2002). *Handbook of Brownian Motion: Facts and Formulae*, 2nd edn. (Birkhauser).

Brennan, M. J. and Schwartz, E. S. (1990). Arbitrage in stock index futures, *Journal of Business* **63**, 1, pp. S7–S31.

Brigo, D., Pallavicini, A., and Torresetti, R. (2007). Calibration of CDO tranches with the dynamical generalized-Poisson loss model, *Risk* **20**, 5, pp. 70–75.

Broadie, M., Chernov, M., and Johannes, M. (2009). Understanding index option returns, *Review of Financial Studies* **22**, 11, pp. 4493–4529.

Cartea, A., Jaimungal, S., and Penalva, J. (2015). *Algorithmic and High-Frequency Trading* (Cambridge University Press, Cambridge, England).

Casassus, J. and Collin-Dufresne, P. (2005). Stochastic convenience yield implied from commodity futures and interest rates, *The Journal of Finance* **60**, 5, pp. 2283–2331.

Chiu, M. C. and Wong, H. Y. (2012). Dynamic cointegrated pairs trading: Time-consistent mean-variance strategies, Tech. rep., working paper.

Cornell, B., Cvitanić, J., and Goukasian, L. (2007). Optimal investing with perceived mispricing, Working paper.

Cox, J. C., Ingersoll, J., and Ross, S. A. (1981). The relation between forward prices and futures prices, *Journal of Financial Economics* **9**, 4, pp. 321–346.

Cox, J. C., Ingersoll, J. E., and Ross, S. A. (1985). A theory of the term structure of interest rates, *Econometrica* **53**, 2, pp. 385–408.

Czichowsky, C., Deutsch, P., Forde, M., and Zhang, H. (2015). Portfolio optimization for an exponential Ornstein-Uhlenbeck model with proportional transaction costs, Working paper.

Dai, M., Zhong, Y., and Kwok, Y. K. (2011). Optimal arbitrage strategies on stock index futures under position limits, *Journal of Futures Markets* **31**, 4, pp. 394–406.

Davis, M. (1997). Option pricing in incomplete markets, in M. Dempster and S. Pliska (eds.), *Mathematics of Derivatives Securities* (Cambridge University Press), pp. 227–254.

Dayanik, S. (2008). Optimal stopping of linear diffusions with random discounting, *Mathematics of Operations Research* **33**, 3, pp. 645–661.

Dayanik, S. and Karatzas, I. (2003). On the optimal stopping problem for one-dimensional diffusions, *Stochastic Processes and their Applications* **107**, 2, pp. 173–212.

d'Halluin, Y., Forsyth, P. A., and Vetzal, K. R. (2005). Robust numerical methods for contingent claims under jump diffusion processes, *IMA Journal of Numerical Analysis* **25**, pp. 87–112.

Ding, X., Giesecke, K., and Tomecek, P. (2009). Time-changed birth processes and multi-name credit derivatives, *Operations Research* **57**, 4, pp. 990–1005.

Driessen, J. (2005). Is default event risk priced in corporate bonds? *Review of Financial Studies* **18**, 1, pp. 165–195.

Duffee, G. R. (1999). Estimating the price of default risk, *Review of Financial Studies* **12**, 1, pp. 197–226.

Duffie, D. and Garleanu, N. (2001). Risk and valuation of collateralized debt obligations, *Financial Analysts Journal* **57**, 1, pp. 41–59.

Duffie, D., Pan, J., and Singleton, K. J. (2000). Transform analysis and asset pricing for affine jump-diffusions, *Econometrica* **68**, 6, pp. 1343–1376.

Dunis, C. L., Laws, J., Middleton, P. W., and Karathanasopoulos, A. (2013). Nonlinear forecasting of the gold miner spread: An application of correlation filters, *Intelligent Systems in Accounting, Finance and Management* **20**, 4, pp. 207–231.

Dynkin, E. and Yushkevich, A. (1969). *Markov Processes: Theorems and Problems* (Plenum Press).

Egami, M., Leung, T., and Yamazaki, K. (2013). Default swap games driven by spectrally negative Lévy processes, *Stochastic Processes and their Applications* **123**, 2, pp. 347–384.

Ekström, E., Lindberg, C., Tysk, J., and Wanntorp, H. (2010). Optimal liquidation of a call spread, *Journal of Applied Probability* **47(2)**, pp. 586–593.

Elliott, R., Van Der Hoek, J., and Malcolm, W. (2005). Pairs trading, *Quantitative Finance* **5**, 3, pp. 271–276.

Engel, C. and Hamilton, J. D. (1989). Long swings in the exchange rate: Are they in the data and do markets know it? Tech. rep., National Bureau of Economic Research.

Engle, R. F. and Granger, C. W. (1987). Co-integration and error correction: representation, estimation, and testing, *Econometrica* **55**, 2, pp. 251–276.

Errais, E., Giesecke, K., and Goldberg, L. R. (2010). Affine point processes and portfolio credit risk, *SIAM Journal on Financial Mathematics* **1**, pp. 642–665.

Evans, L. C. (1998). *Partial Differential Equations* (AMS Graduate Studies in Mathematics).

Ewald, C.-O. and Wang, W.-K. (2010). Irreversible investment with Cox-Ingersoll-Ross type mean reversion, *Mathematical Social Sciences* **59**, 3, pp. 314–318.

Feller, W. (1951). Two singular diffusion problems, *The Annals of Mathematics* **54**, 1, pp. 173–182.

Föllmer, H. and Schied, A. (2002). Convex measures of risk and trading constraints, *Finance Stoch.* **6**, 4, pp. 429–447.

Föllmer, H. and Schied, A. (2004). *Stochastic Finance: An Introduction in Discrete Time*, 2nd edn., De Gruyter Studies in Mathematics (Walter de Gruyter).

Föllmer, H. and Schweizer, M. (1990). Hedging of contingent claims under incomplete information, in M. Davis and R. Elliot (eds.), *Applied Stochastic Analysis, Stochastics Monographs*, Vol. 5 (Gordon and Breach, London/New York), pp. 389 – 414.

Forsyth, P. A., Kennedy, J. S., Tse, S. T., and Windcliff, H. (2012). Optimal trade execution: A mean-quadratic-variation approach, *Journal of Economic Dynamics and Control* **36**, pp. 1971–1991.

Frei, C. and Westray, N. (2013). Optimal execution of a VWAP order: A stochastic control approach, *Mathematical Finance* **25**, 3, pp. 612–639.

Fritelli, M. (2000). The minimal entropy martingale measure and the valuation problem in incomplete markets, *Mathematical Finance* **10**, pp. 39–52.

Fujiwara, T. and Miyahara, Y. (2003). The minimal entropy martingale measures for geometric Lévy processes, *Finance Stoch.* **7(4)**, pp. 509–531.

Gatev, E., Goetzmann, W., and Rouwenhorst, K. (2006). Pairs trading: Performance of a relative-value arbitrage rule, *Review of Financial Studies* **19**, 3, pp. 797–827.

Glowinski, R. (1984). *Numerical Methods for Nonlinear Variational Problems* (Springer-Verlag, New York).

Göing-Jaeschke, A. and Yor, M. (2003). A survey and some generalizations of Bessel processes, *Bernoulli* **9**, 2, pp. 313–349.

Gropp, J. (2004). Mean reversion of industry stock returns in the US, 1926–1998, *Journal of Empirical Finance* **11**, 4, pp. 537–551.

Grübichler, A. and Longstaff, F. (1996). Valuing futures and options on volatility, *Journal of Banking and Finance* **20**, 6, pp. 985–1001.

Hamilton, J. D. (1994). *Time Series Analysis*, Vol. 2 (Princeton university press Princeton).

Henderson, V. and Hobson, D. (2011). Optimal liquidation of derivative portfolios, *Mathematical Finance* **21**, 3, pp. 365–382.

Henderson, V., Hobson, D., Howison, S., and T.Kluge (2005). A comparison of q-optimal option prices in a stochastic volatility model with correlation, *Review of Derivatives Research* **8**, pp. 5–25.

Heston, S. L. (1993). A closed form solution for options with stochastic volatility with applications to bond and currency options, *Review of Financial Studies* **6**, pp. 327–343.

Hobson, D. (2004). Stochastic volatility models, correlation, and the q-optimal measure, *Mathematical Finance* **14**, 4, pp. 537–556.

İlhan, A., Jonsson, M., and Sircar, R. (2005). Optimal investment with derivative securities, *Finance and Stochastics* **9**, 4, pp. 585–595.

Itō, K. and McKean, H. (1965). *Diffusion processes and their sample paths* (Springer Verlag).

Jarrow, R. A., Lando, D., and Yu, F. (2005). Default risk and diversification: Theory and empirical implications, *Mathematical Finance* **15**, 1, pp. 1–26.

Jurek, J. W. and Yang, H. (2007). Dynamic portfolio selection in arbitrage, Working Paper, Princeton University.

Kanamura, T., Rachev, S. T., and Fabozzi, F. J. (2010). A profit model for spread trading with an application to energy futures, *The Journal of Trading* **5**, 1, pp. 48–62.

Karatzas, I. and Shreve, S. (1998). *Methods of Mathematical Finance* (Springer, New York).

Karatzas, I. and Shreve, S. E. (1991). *Brownian Motion and Stochastic Calculus* (Springer, New York).

Karlin, S. and Taylor, H. M. (1981). *A Second Course in Stochastic Processes*, Vol. 2 (Academic Press).

Kladivko, K. (2007). Maximum likelihood estimation of the Cox-Ingersoll-Ross process: the MATLAB implementation, Technical Computing Prague.

Kong, H. T. and Zhang, Q. (2010). An optimal trading rule of a mean-reverting asset, *Discrete and Continuous Dynamical Systems* **14**, 4, pp. 1403 – 1417.

Larsen, K. S. and Sørensen, M. (2007). Diffusion models for exchange rates in a target zone, *Mathematical Finance* **17**, 2, pp. 285–306.

Lebedev, N. (1972). *Special Functions & Their Applications* (Dover Publications).

Leung, T. and Li, X. (2015). Optimal mean reversion trading with transaction costs and stop-loss exit, *International Journal of Theoretical & Applied Finance* **18**, 3, p. 15500.

Leung, T., Li, X., and Wang, Z. (2014). Optimal starting–stopping and switching of a CIR process with fixed costs, *Risk and Decision Analysis* **5**, 2, pp. 149–161.

Leung, T., Li, X., and Wang, Z. (2015). Optimal multiple trading times under the exponential OU model with transaction costs, *Stochastic Models* **31**, 4.

Leung, T. and Liu, P. (2013). An optimal timing approach to option portfolio risk management, in J. Batten, P. MacKay, and N. Wagner (eds.), *Advances in Financial Risk Management: Corporates, Intermediaries, and Portfolios* (Palgrave Macmillan), pp. 391–403.

Leung, T. and Ludkovski, M. (2011). Optimal timing to purchase options, *SIAM Journal on Financial Mathematics* **2**, 1, pp. 768–793.

Leung, T. and Ludkovski, M. (2012). Accounting for risk aversion in derivatives purchase timing, *Mathematics & Financial Economics* **6**, 4, pp. 363–386.

Leung, T. and Shirai, Y. (2015). Optimal derivative liquidation timing under path-dependent risk penalties, *Journal of Financial Engineering* **2**, 1, p. 1550004.

Leung, T., Sircar, R., and Zariphopoulou, T. (2012). Forward indifference valuation of American options, *Stochastics: An International Journal of Probability and Stochastic Processes* **84**, 5-6, pp. 741–770.

Leung, T. and Yamazaki, K. (2013). American step-up and step-down credit default swaps under Lévy models, *Quantitative Finance* **13**, 1, pp. 137–157.

Lin, Y.-N. and Chang, C.-H. (2009). Vix option pricing, *Journal of Futures Markets* **29**, 6, pp. 523–543.

Longstaff, F. A. and Rajan, A. (2008). An empirical analysis of the pricing of collateralized debt obligations, *The Journal of Finance* **63**, 2, pp. 529–563.

Lopatin, A. and Misirpashaev, T. (2008). Two-dimensional Markovian model for dynamics of aggregate credit loss, *Advances in Econometrics* **22**, pp. 243–274.

Lorenz, J. and Almgren, R. (2011). Mean-variance optimal adaptive execution, *Applied Mathematical Finance* **18**, 5, pp. 395–422.

Malliaropulos, D. and Priestley, R. (1999). Mean reversion in Southeast Asian stock markets, *Journal of Empirical Finance* **6**, 4, pp. 355–384.

Menaldi, J., Robin, M., and Sun, M. (1996). Optimal starting-stopping problems for Markov-Feller processes, *Stochastics: An International Journal of Probability and Stochastic Processes* **56**, 1-2, pp. 17–32.

Mencía, J. and Sentana, E. (2013). Valuation of VIX derivatives, *Journal of Financial Economics* **108**, 2, pp. 367–391.

Metcalf, G. E. and Hassett, K. A. (1995). Investment under alternative return assumptions comparing random walks and mean reversion, *Journal of Economic Dynamics and Control* **19**, 8, pp. 1471–1488.

Mortensen, A. (2006). Semi-analytical valuation of basket credit derivatives in intensity-based models, *Journal of Derivatives* **13**, 4, pp. 8–26.

Øksendal, B. (2003). *Stochastic Differential Equations: an Introduction with Applications* (Springer).

Oksendal, B. and Sulem, A. (2005). *Applied Stochastic Control of Jump Diffusions* (Springer).

Poterba, J. M. and Summers, L. H. (1988). Mean reversion in stock prices: Evidence and implications, *Journal of Financial Economics* **22**, 1, pp. 27–59.

Ribeiro, D. R. and Hodges, S. D. (2004). A two-factor model for commodity prices and futures valuation, EFMA 2004 Basel Meetings Paper.

Rockafellar, R. and Uryasev, S. (2000). Optimization of conditional value-at-risk, *The Journal of Risk* **2**, 3, pp. 21–41.

Rogers, L. and Williams, D. (2000). *Diffusions, Markov Processes and Martingales*, Vol. 2, 2nd edn. (Cambridge University Press, UK).

Rogers, L. C. G. and Singh, S. (2010). The cost of illiquidity and its effects on hedging, *Mathematical Finance* **20**, 4, pp. 597–615.

Schied, A. and Schöneborn, T. (2009). Risk aversion and the dynamics of optimal liquidation strategies in illiquid markets, *Finance and Stochastics* **13**, 2, pp. 181–204.

Schönbucher, P. J. (2003). *Credit Derivatives Pricing Models: Models, Pricing, Implementation* (Wiley Finance).

Schwartz, E. (1997). The stochastic behavior of commodity prices: Implications for valuation and hedging, *The Journal of Finance* **52**, 3, pp. 923–973.

Sircar, R. and Papanicolaou, A. (2014). A regime-switching Heston model for VIX and S&P 500 implied volatilities, *Quantitative Finance* **14**, 10, pp. 1811–1827.

Song, Q., Yin, G., and Zhang, Q. (2009). Stochastic optimization methods for buying-low-and-selling-high strategies, *Stochastic Analysis and Applications* **27**, 3, pp. 523–542.

Song, Q. and Zhang, Q. (2013). An optimal pairs-trading rule, *Automatica* **49**, 10, pp. 3007–3014.

Sun, M. (1992). Nested variational inequalities and related optimal starting-stopping problems, *Journal of Applied Probability* **29**, 1, pp. 104–115.

Tourin, A. and Yan, R. (2013). Dynamic pairs trading using the stochastic control approach, *Journal of Economic Dynamics and Control* **37**, 10, pp. 1972–1981.

Triantafyllopoulos, K. and Montana, G. (2011). Dynamic modeling of mean-reverting spreads for statistical arbitrage, *Computational Management Science* **8**, 1-2, pp. 23–49.

Tsay, R. S. (2005). *Analysis of Financial Time Series*, Vol. 543 (John Wiley & Sons).

Vidyamurthy, G. (2004). *Pairs Trading: Quantitative Methods and Analysis* (Wiley).

Zervos, M., Johnson, T., and Alazemi, F. (2013). Buy-low and sell-high investment strategies, *Mathematical Finance* **23**, 3, pp. 560–578.

Zhang, H. and Zhang, Q. (2008). Trading a mean-reverting asset: Buy low and sell high, *Automatica* **44**, 6, pp. 1511–1518.

Zhang, J. E. and Zhu, Y. (2006). VIX futures, *Journal of Futures Markets* **26**, 6, pp. 521–531.

Index

Printed in the United States
By Bookmasters